COMPLEXITY

AND EMERGENCE

Proceedings of the Annual Meeting of the International Academy of the Philosophy of Science

COMPLEXITY

AND EMERGENCE

Bergamo, Italy — 9 – 13 May 2001

Editors

Evandro Agazzi & Luisa Montecucco

University of Genoa, Italy

Published by

World Scientific Publishing Co. Pte. Ltd.

P O Box 128, Farrer Road, Singapore 912805

USA office: Suite 202, 1060 Main Street, River Edge, NJ 07661

UK office: 57 Shelton Street, Covent Garden, London WC2H 9HE

British Library Cataloguing-in-Publication Data
A catalogue record for this book is available from the British Library.

ISBN 981-238-158-9

This book is printed on acid-free paper.

Printed in Singapore by Uto-Print

Contents

PART III   THE EMERGENCE OF THE MIND

Introduction

Complexity has become a dominant concept in contemporary science, especially as a consequence of the increasing impact of the perspectives of general systems-theory. And its fruitfulness essentially consists in the fact that it has rendered possible a deeper understanding of the functional and structural interrelations existing within a given system and between this system and surrounding systems. The advantage of this point of view is that it offers conceptual tools for a more adequate understanding of the *novelties* that complex systems evince with regard to their constituent parts, novelties that are often called *emergent* and that give a rather precise sense to the old maxim that "the whole is more than the sum of its parts". This "more", however, can be interpreted according to two contrary approaches. The first consists in maintaining that the knowledge of the parts and their properties is sufficient to account for the new properties of the whole. Since these are *deterministically* entailed by the properties of the parts, they *result* from them and, in this sense, are *reducible* to them. The second consists in maintaining that the properties of the parts are *necessary but not sufficient* to account for the properties of the whole, since the latter are further dependent on the occurrence of particular conditions not inscribed in the properties of the parts, and whose contingent configuration could have led to a totally different *result* if they had been only slightly different. In this sense the second approach relies on an *indeterministic* view of the dynamics of natural phenomena (which is also extended to phenomena studied in disciplines different from the natural sciences).

This fact explains why the thematic of complexity has become more important in the natural sciences as a consequence of the obsolescence of the classic deterministic view and of some of its most characteristic corollaries. The classic view is well portrayed in the famous statement of Laplace: a superior Intelligence that could know at a given instant the exact state of the universe (that is, the positions and velocities of all its material points) and also know all the natural laws regulating their motion, and that in addition could also calculate the exact solutions of the differential equations expressing these laws, would be able to predict exactly the state of the universe (that is, the position and velocity of every material point in it) at any future instant. Since humans are far from being endowed with such exceptional cognitive powers, they must compensate their *ignorance* by resorting to the calculus of *probabilities*, thanks to which they are able, starting from a good *approximation* in their knowledge of the initial conditions, to predict with an equally good approximation the knowledge of future events. And this because a "small" change in the initial conditions was believed to lead to a "small" change in the deterministically following final state. This tacit postulate is known as the condition of *linearity* for the physical phenomena; and it is well-known that already with Poincaré this postulate was seriously challenged in his study of the famous "three-body problem", in which the application of the most deterministic of the laws of physics (Newtonian gravitation) was shown to be insufficient to grant the expected exact prediction even of the behavior of such a simple system as that constituted by three material bodies. The awareness of the *non-linearity* of the great majority of physical effects is now one of the most generally recognized features of scientific phenomena, and appears as a direct consequence of their *complexity*.

A strong aspect of the classical perspective seemed to be that it was the best suited to account for the *order* and harmony of the universe. If this is regulated by eternal, immutable and strictly deterministic laws, it is sufficient to admit that a certain wise disposition of the initial state of the universe was posited (perhaps by a divine intelligence, as Newton himself had suggested), in order to be sure that this "cosmos" (that is, ordered totality) will persist eternally and be intrinsically explorable by humans. Within classical physics, however, this view soon appeared hardly tenable. When the second principle of thermodynamics expressed the "irreversibility" of thermal phenomena, the problem became that of reconciling this feature with the time-reversibility of all the mechanical laws.

The ingenious solution elaborated by certain famous physicists of the 19th century was at the same time reductionist and probabilistic: heat and temperature are only macroscopic magnitudes corresponding to the mechanical kinetic energy of the molecules constituting matter, which are subject to continuous agitation and collisions regulated by classical reversible mechanical laws. In this interpretation something such as the "irreversible" passage of heat from a body at a high temperature to a body at a lower temperature was explained to be not really irreversible, but only endowed with a very high *probability*. Therefore, reversibility in thermodynamics was not theoretically excluded, but shown to be so highly improbable a process that it should be *practically* excluded. In the technical elaborations of this point two salient aspects are worth mentioning: *disorder* is much more probable than *order*, such that the natural tendency of the dynamic of a closed system is towards an increasing state of disorder of its constituent parts. In terms of energy, this was expressed as a law of the *degradation of energy* according to which the evolution of a closed system is toward a state of *thermal death*, in which all forms of energy have been reduced to the lowest form, i.e., to heat. Since the universe was considered a closed system, this was the famous thesis of the thermal death of the universe. As a conclusion of this story, the deterministic view, once it had been applied to the understanding of complex systems and of real processes, appeared to entail a transition from order to disorder and from stability to death.

More appealing was the perspective elaborated in a much more intuitive and less "rigorous" way by biology. Determinism has always been rather alien to the conceptual space of biology, and this precisely because any living organism is in itself such a complex entity that an exact prediction of its behaviour is almost meaningless. The most we can do is to formulate *probabilistic* predictions based on the empirical ascertainment of *relative frequencies*. This general attitude was greatly reinforced by the appearance of *evolutionistic theories* in the 19th century. Independently of the differences existing among these theories concerning the postulation of the "mechanisms" of evolution, all of them agreed on the admission (that was also supported by empirical evidence) of the unpredictable appearance (and extinction) of living species. In particular, *novelties* constituted by the appearance of new species were accounted for by an *improbable* encounter of contingent situations.

According to the Darwinian theory of natural selection (but, to a minor degree, also according to other theories of evolution), the concurrent presence of certain contingent "exceptional" features in a few individuals of a given species—that also contingently happened to make them more fit vis-à-vis their environment and better able to survive in it, and which also happened to be inherited by their descendants—would gradually lead to the extinction of the offspring of the "normal" individuals and to the consolidation of the new species endowed with the "favorable" characteristics which initially occurred "by chance". According to this view, the "progress" of the biosphere, with the multiplication of new species, was produced by fortuitous, highly improbable and unpredictable events; and this view was at variance with the view of classical physics, according to which the "evolution" of the physical world is toward the "most probable" state of equilibrium, imposed by the deterministic laws of nature.

But there is more: the said "progress" of the biosphere does not only consist in the fact that the number of living species has increased prodigiously, but also in the fact that it has consisted in the transition from less *complex* to more *complex* forms of life, such that we could define in general as evolutionistic any theory according to which the total of presently existing species *are derived* from earlier species that were *less numerous and less complex*. The notion of complexity implicit in this definition is still rather vague, but can already be seen as an increase of *order*, of a higher level *coordination* that, in particular, leads to the appearance of *new* properties and functions. What is noteworthy is that, in this perspective, order and coordination come out of disorder, and that they are not the "result" of some deterministic mechanisms hidden in the initial state, but of a contingent succession of fortuitous events.

As is well known, this view has conquered by now several sectors of the physical sciences as well: order by fluctuation, deterministic chaos, and non-linear processes are among the domains in which research is currently being carried out; and, in particular, *cosmology* has extended the evolutionistic point of view to the understanding of the universe as a whole. The universe has had an "origin"; the fundamental physical forces were not all present and active at the initial moments of its coming into being, only very few of the chemical elements have existed for a long time, and, similarly to living species, the different kinds of stars have a definite lifespan. Thus even a very "small" difference in the value of certain physical constants would have led to a drastically different history or "evolution" of the universe in which, in particular, the conditions for the existence of life would not have occurred. In brief, the transition from simplicity to complexity, and the related emergence of novel unpredictable aspects of the universe, is now considered as a well established fact.

Despite all this, a sufficiently clear determination of the very notions of complexity and emergence is still to be elaborated, and this is not surprising since these concepts are bound, on the one hand, to a common meaning and, on the other hand, are not treated univocally in different scientific contexts. Thus a clarification of these notions appears warranted, and must be attempted through an interdisciplinary approach that cannot pretend to offer an indisputable "definition" of them, but should more modestly aim at providing a careful analysis of certain fundamental meanings that can be considered as "transdisciplinary", along with the clarification of other "specific" meanings that are typical of different disciplinary contexts. The task of proposing such an analysis has been pursed in an annual meeting of the International Academy for the Philosophy of Science with the name "Complexity and Emergence", held in Bergamo (Italy) from May 9-13, 2001, and organized in collaboration with Bergamo University. The present volume contains most of the invited papers of this meeting.

The first part of the volume, devoted to a general discussion of these notions, opens with a paper by **Evandro Agazzi** ('What is complexity?'), in which a definition of complexity is looked for by means of a contraposition with its semantic contrary "simplicity". After having recognized that simple and complex are determinable only in relation to a given context (so that what is simple from one point of view can be complex from another), and having stressed that these notions apply to specific "wholes", a distinction is drawn between "analytic" and "synthetic" simplicity, based on the consideration of the "internal" and "external" relations a given "whole" may have with other systems. This distinction has obvious counterparts in the notion of complexity. These considerations allow a clarification of the links existing between complexity and *order* or *structure*. This analysis permits us to admit a legitimate status to *reductionism* under particular conditions, and to see that emergence is not a consequence of all instances of complexity. Real emergence occurs when the analysis of the "simples" constituting a "complex" whole is not *sufficient* for explaining the *new attributes* characteristic of the whole, and when those new attributes are not used in the characterization of the whole's "simple" constituents.

The mention of levels and perspectives alluded to in the foregoing paper is developed in an original way in the paper by **Hans Lenk** and **Achim Stephan** ('Levels and types of complexity and emergence'). The approach of this paper is methodological and meta-theoretical, and essentially consists in outlining the importance of considering the different meta-levels on which complexity can be investigated. The core of this approach consists in recognizing the role played by *interpretation* and, more precisely, by *schema-interpretation*. On the basis of this role, different kinds of "emergentism" (weak emergentism, synchronic emergentism, diachronic emergentism) are surveyed and analytically discussed, and a balanced judgment is put forward regarding the pros and cons of emergence. In conclusion, no less important than the recognition of the ontological complexity we encounter in the world, is understanding the complexity of the descriptions and interpretations we develop in order to account for this ontological complexity. Owing to this fact different notions of complexity and emergence appear in the different sciences and in the philosophy of science.

The paper by **C. Ulises Moulines** ('Formal metatheoretical criteria of complexity and emergence') is also of an epistemological nature. After having noted that complexity C and emergence E can be considered, on an ontological level, as ordering relations between entities, the basic intuitive characteristics of these relations are outlined: C is reflexive, transitive and non-connected; and E is asymmetric, transitive and non-connected. It seems also intuitively obvious that an increase in complexity is needed for emergence. This intuitive ontological consideration, however, is open to question in several cases. The proposal of the paper is therefore that of considering complexity and emergence as relations between theories, since this move allows the application of several precise instruments elaborated in the philosophy of science. In particular, Moulines resorts to concepts and methods of the so-called "structuralist" view of scientific theories, and tries to show that a structural comparison of the "models" of two theories regarding different domains of application can provide useful criteria for comparing their complexity. Considering intertheoretical reduction, he notes that this has an "ontological" side and a "nomological" side: when nomological reduction is not obtained we can already speak of emergence: if in the presence of ontological reduction, "weak emergence", and otherwise "strong emergence". This approach has the advantage of superseding the overly hasty intuitive idea that increase in complexity itself entails emergence.

The paper of **Bernulf Kanitscheider** ('Beyond reductionism and holism. The approach of synergetics') also relies upon a particular metatheoretical framework, that is, the theory of synergetics. The merit of this approach is that of overcoming the sterile and almost ideological controversy that for many decades has divided the partisans of reductionism and holism. This is done through the elaboration of a general kind of consideration, operating at a meta-level, for the treatment of the autonomous organization of complex systems with arbitrary material properties. It is not yet clear whether this theory is really able to construct a comprehensive scheme of morphogenetics but, if this were the case, it would offer a new bridge between the natural, social and mental sciences. The binding principle would consist in the unitary description of cooperative emergent phenomena. The effort of synergetics is that of linking microlevels and macrolevels on the basis of three principles: parameters of order, the principle of enslavement, and circular causality. The paper offers a discussion of the concrete conditions for the applicability of this pattern.

The ambition to attain a "precise" concept is often equated with the possibility of giving it a mathematical representation that, in particular, could make it "measurable". This has also happened in the case of complexity, the most famous mathematical formulation of which was introduced by Kolmogorov. **Jesús Mosterín** ('Kolmogorov complexity') offers a clear presentation of this approach, whose starting point consists in a transition from an ontological to a linguistic kind of consideration: degree of complexity is judged by considering the *description* of an entity and, more precisely, consists in the measure of the size of the shortest description of this entity. Such a shortest description is seen as the shortest program that can generate the whole description in a universal Turing machine. For this reason we can say that Kolmogorov's is an "algorithmic theory of complexity", since this notion is equated with the possibility of finding the shortest program that can generate the description according to a universal method of computation. In fact we can say that a particular program is the shortest when it is no longer possible to "compress" it in the sense of information theory (in this sense complexity amounts to a resistance to compression), and the universality of a Turing machine makes the measure of complexity independent of any particular computer in which the program could be made operational; such a computer can always be equivalently replaced by a universal Turing machine. The function k expressing Kolmogorov complexity is not itself computable, but has computable approximations; and this is sufficient for any practical purpose. The idea that any object can undergo a mathematical description may sound somewhat pretentious, but in the paper it is explained that this does not mean that the object must be analyzed in terms of mathematical concepts or functions, but simply that it is possible to provide a mathematical "encoding" of any description. As to the notion of compressibility, it is noted that the axiomatic method, for example, is already a familiar tool for the compression of information, since all the theorems of an axiomatized theory are, so to speak, compressed in the axioms from which they can be derived by the application of such algorithms as logical rules and mathematical computations. Kolmogorov's approach (1965) has not been exempt from criticism, and the paper mentions certain developments and improvements that have been proposed for superseding some of its limitations by several mathematicians (such as Chaitin, Martin-Löf, Levin, Gacs, Loveland). In particular, in order to overcome the limitation of considering complexity as essentially related to randomness, the position of Bennet is presented in which the "logical depth" of an object is taken into consideration and defined as the time (that is, the number of steps) required to generate it from its maximally compressed description.

Whereas mathematical technicalities are avoided in Mosterín's paper, they are widely used in the paper by **Jean Petitot** ('Modèles de Structures Émergentes dans les Systèmes Complexes'), and this because the author wants to show concretely how the emergence of "forms" or "patterns" can be accounted for by applying the ideas of R. Thom, who was the first to propose a mathematical theory of morphology and morphogenesis: local dynamic processes interact and give rise to instabilities that produce new forms. Petitot offers a detailed analysis of two significant examples, one regarding chemical processes (according to a model proposed by Turing and elaborated by Coullet), and one regarding the neurosciences (namely neuronal pattern-recognition). The in-depth mathematical discussion cannot be rendered in this short summary.

The second part of the volume contains a few papers concerning complexity and emergence in natural science, or, more precisely, in physics and biology. As to physics, reductionism has been the prevailing position until the end of the 19th century, owing to the almost absolute value attributed to mechanics, which was considered the fundamental science capable of offering the key to the interpretation and explanation of all natural phenomena. After the decline of this tenet, however, a similar problem has often been discussed. Since quantum mechanics is now considered as the theory concerned with the truly basic realities of physics, it is a theoretical challenge to provide correct accounts in quantum mechanical terms of the visible or macroscopic features of the world that used to be considered described and explained by classical physics. This task is not necessarily equated with a "reduction" to quantum physics of classical physics, but at least as a possibility of harmonizing certain basics features of these two physical theories, which are otherwise seriously at variance. This is, essentially, the aim pursued in the papers of Omnés and Cordero.

Roland Omnés ('Emergence in physics: the case of classical physics'), after an historical discussion surveying the fundamental factors of contraposition between classical and quantum physics, maintains that a renewal of the interpretation of quantum mechanics, through the introduction of new notions such as decoherence, semi-classical physics, consistent histories and standardized quantum logic, makes it plausible to speak of an "emergence" of classical physics from quantum mechanics. He mentions a list of alleged incompatibilities between quantum mechanics and classical physics, and shows how these "gaps" could be settled little by little through the application of these new concepts.

The proposal of **Alberto Cordero** ('Classical properties in a quantum-mechanical world') is similar, for he mentions a list of "desiderata" whose fulfillment would open the way to a satisfactory unification of classical and quantum physics. These desiderata can be summarized as the aim of showing that a classical system is a large composite quantum-system, and that its differences with regard to quantum properties are due to its complexity. The paper considers the approach proposed by Ghirardi, Rimini and Weber (GRW) and analyses to what extent a consistent application of this approach can cope with the mentioned desiderata. Certain shortcomings are recognized as being due to the peculiarities of the GRW program, but other shortcomings also affect in general different approaches to a unification theory of physics. The author maintains that the reason for such shortcomings can be traced back to certain commonly admitted presuppositions about the status of Boolean structures in physics, and suggests that relaxing these presuppositions can open the way to meeting the original desiderata.

Jacques Ricard ('Reduction Integration, Emergene and Complexity in Biological Networks') starts by proposing definitions of the notions of reduction, integration, emergence and complexity that are particularly significant in biology, and lays stress on the importance of the history of a complex system for the study of its characteristics and dynamics. The rest of the paper is essentially devoted to the critical discussion of one significant example, that of "biological networks". The tools used in this discussion are taken from the probability calculus and information theory, and are applied with appropriate technical details that cannot be summarized here. We can mention, however, that an important part of this discussion shows that the classical Shannon information theory is not suited to coping with phenomena where only correlation, but not physical interaction, is taken into account (and this is typical of biological systems). He then shows that departure from quasi-equilibrium results in the emergence of information: in this case the system does not simply act as a channel for information, but as a source of information of its own. The same can happen when a system, under steady-state conditions, departs from equilibrium.

The third part concerns a question that has been debated for centuries in philosophy and the sciences: the status of the mind with regard to matter, especially with regard to such "mental" characteristics as meaning, intentionality and consciousness, which seem hardly applicable or even conceivable in the domain of matter, but which must nevertheless be related to matter since those beings that are endowed with mind are also material. The most reasonable strategy in discussing this intriguing issue seems to be that of bringing these two domains into as close contact as possible without pretending a priori either to reduce them to a fictitious unity, or to keep them drastically separated.

With regard to this fact, the approach proposed by **F. Tito Arecchi** ('Complexity and the emergence of meaning. Toward a semiophysics') is particularly interesting, for it shows a way to consider meaning as something that is not out of conceptual reach within a physical way of thinking (that is, from the point of view of physics). The core of his presentation is prepared by the author through the analysis of what makes particular aspects of reality susceptible of scientific treatment, with the linguistic consequences entailed by this. This discourse is supported by several technical and mathematical considerations. The topic of complexity is introduced by distinguishing closed systems and open systems, and pure complication from complexity in a proper sense. The third part of the paper concerns understanding, by means of a physical description, how external stimuli are transformed into sensorial perceptions. In this way a bridge is offered between physics and the neurosciences (a "neurophysics of perception", as the author calls it). This longer paper, which summarizes original research by the author over many years, requires a certain familiarity with mathematics and physics to be fully appreciated, though it also contains several general reflections that can be of use to non-specialists.

Whereas Arecchi has preferred to discuss the properties of the "brain" without entering into the complicated issue of the "mind", **Mario Casartelli**

('Complexity and the emergence of intentionality: some misconceptions'), who is also a professional physicist, explicitly addresses the mind-body problem, and focuses his critical examination on certain claims of the strong artificial intelligence (SAI) thesis. He challenges first the strictly behavioristic approach that almost universally characterizes, as a tacit presupposition, debates regarding this thesis. The crucial point in these debates has concerned the role of intentionality, and efforts have been made to maintain that we can dispense with it. The author notes that even in the case of a totally "deterministic" machine, this presupposes an "external" program that was "intentionally" predisposed for its functioning; and this external start is also unavoidable in the (still hypothetical and speculative) case of self-programming computers, in which pure complexity cannot account for the initial program that makes them self-programming. Therefore, he prefers to speak of "embodiment" rather than of emergence, since in such a way we can seriously take into account that the brilliant performances of intelligent machines depend on their being projected and programmed in a world where intentionality is a precondition for their understanding. The same kind of considerations also hold regarding the "eliminative" interpretations of the neurosciences. Complexity is not sufficient as a "cause" of consciousness.

A general concern of people who advocate emergence is often that of avoiding reductionism. In the debate on the mind-body problem a rather special concept has been presented with a similar aim, that of "supervenience". **Luisa Montecucco** ('Can supervenience save the mental?') recognizes that the theories of supervenience and emergence are similar in the sense that both hold that the family of mental properties are clearly dependent on a qualitatively distinct family of physical properties without necessarily being reducible to them; and, in order to explore the concept of supervenience, she makes a few remarks concerning the different contextual development of the two theories and their different focal points with respect to physicalism without reductionism. Specific references are made to D. Davidson (for having introduced the term "supervenience" to characterize a dependency relation which may hold between mental and physical properties even if there are no laws connecting them), and to J. Kim (for having made this concept more precise through his inquiry on varying strengths of supervenience intended to catch the intuitive ideas of dependence without reducibility). With the addition of property covariance (something indiscernible in subvenient properties must be indiscernible in supervenient properties), these are constitutive ideas of supervenience as well as of emergence. Given this conceptual framework, the paper takes a critical attitude regarding the theory of supervenience as a suitable solution to the mind-body problem, as well as regarding its effectiveness in avoiding the reducibility of the mental to the physical.

The mention of "soul" has not been made in the papers considered up to now, perhaps because it has a too "metaphysical" flavor. A professional biochemist, however, explicitly uses it in the title of his contribution, **Giuseppe Del Re** ('From complexity to the separate soul'). In the first part of the paper certain preliminary clarifications are presented. One regards the distinction between "process complexity" and "systems complexity" (the complexity resulting from the relations between the whole and its parts), and only the second is investigated by the author. In addition, certain specifically philosophical notions are considered relevant to this discussion, namely the Aristotelian concept of potency and the idea of "levels of complexity" (investigated along the lines of N. Hartmann's ontology). The discussion is deepened by an analysis of the concepts of structure, organization and information. The second part discusses the nature of life, making use of the previously examined concepts in the interpretation of living beings. In this context the "soul" is defined as the organizing principle of living beings, that is, as something that is fully "natural" and corresponds to their specific status of complexity and organization—a new level presupposing the inferior levels and bringing them to a superior organization. In this spirit the psyche also receives a non-substantialistic interpretation: it is a being as characterized by certain properties emerging from the interactions of the elementary objects belonging to the level of complexity defined by neurophysiology. The philosophically challenging problem is whether it is also possible to make the traditional notion of the substantiality of the soul (that is, of its existence in itself independently of its being the organizing principle of a living being) compatible with a scientific view. The author substantiates this possibility via an analysis of the position of Thomas Aquinas vis-à-vis existing scientific and technological artifacts. The relation between information and complexity is considered as an opening toward also admitting a "spiritual" level of reality which is not at variance, but in keeping with, the other levels usually investigated by the sciences.

Special thanks go to the International Academy for the Philosophy of Science and to the University of Bergamo, with the support of the Unione degli Industriali della Provincia di Bergamo, for having made possible the conference "Complexity and Emergence" as well as the publication of this book.

Evandro Agazzi

Luisa Montecucco

PART I

THE NOTIONS OF COMPLEXITY AND EMERGENCE

1.   What is Complexity?

Evandro Agazzi

Dept. of Philosophy, University of Genoa, Italy

1. Complexity and simplicity

The notion of complexity has no neatly characterised meaning in common discourse, for sometimes "complex" is understood to be synonymous with "difficult", while in more rigorous contexts it is understood as the contrary of "simple". The second meaning is the one we shall consider here, particularly because it is the one normally applied in science, and the one used in recent epistemological debates about complexity. We can consider 'complex' as a primitive notion, in the sense that its meaning cannot be codified by means of an explicit definition in terms of other, previously understood, concepts. In common discourse we rely upon a spontaneous and implicit understanding of this notion that can be clarified by its contextual use and also be made more explicit by its being related to a specific contrary notion. This is common to all primitive notions: for example 'being' cannot be defined in terms of more primitive concepts, but is "semantised" as the opposite of 'nonbeing'; 'same' cannot really be defined, but its meaning is understood by relating it to its contrary, 'different'; the concept 'part' cannot be defined, but its meaning results from the opposition it has with 'whole'; and so on.

In the case of 'complex' we have just seen that it can be understood as synonymous with 'difficult', in which case it is semantised as the contrary of 'easy'; otherwise it is semantised as the contrary of 'simple'. In conclusion, we can say that 'complexity' will here be understood as the contrary of 'simplicity'. The reasonableness of this characterisation is easily confirmed by the fact that no one would maintain that something simple is (at the same time and in the same respect) complex, and vice versa.

Let us note the reason why a characterisation through reference to a contrary is significant and semantically useful, while pure negation would not provide any definite meaning. In fact, to say of something that it is not, for example, a horse, would not suggest that it be a dog rather than a stone, the square root of a number, Hamlet, or God, since all of these entities, and an infinite number of others, belong to the class of non-horse. On the contrary, primitive notions semantised through an opposition are informative because they clarify one another by being inscribed within a common "semantic field", such that their mutual exclusion plays the role of a symmetric meaning-dependence.

The fact that a primitive notion is semantised through an opposition does not prevent its being susceptible of further clarification depending on the context in which it occurs. For example, in Euclid's *Elements* the notion of a point is clarified by means of a negation ("Point is that which has no parts"); but at the same time a complete understanding of its meaning comes from a consideration of the whole set of postulates in which the notion occurs along with other equally primitive notions, each of which receives in such a way a kind of "contextual definition". Moreover, a given primitive notion can show certain features that, though not amounting to a real definition, can provide *criteria* for its applicability (e.g., if 'same' is semantised as the contrary of 'different', a criterion of sameness in geometry that spontaneously follows is that two figures are the same if they can be superposed).

Considering 'simplicity', we must note that it would be inadequate to consider simples on a purely geometrical or physicalistic ground, that is, as entities that cannot be further divided into parts. The more precise condition would be that a simple not be endowed with *any* general properties, because if it were it would obviously evince a certain complexity. A genuine simple should be nonanalysable also from a conceptual point of view, and be so peculiar that it would not be describable but only "nameable" (as already remarked by Plato with regard to the primitive constituents of reality).

This aspect of absolute simplicity, however, is so paradoxical that it must be more closely scrutinised and refined. In fact, knowing an entity and speaking of it entail knowing and saying "what it is", and this necessarily amounts to ascribing to it certain general attributes that it exemplifies, which could be also attributes of other entities. In other words, any piece of reality we know is necessarily a "connoted" reality, i.e. something endowed with certain properties or attributes, and we can know this reality only by knowing these attributes. Medieval philosophers were aware of this fact when they said that, though substance is different from its accidents (i.e., attributes), we can know the substance only *through* its accidents.

Let us note that this has direct consequences also for the issue of existence: that something definite exists can only be ascertained by checking whether or not the set of attributes this something ought to posses are actually present, that is, whether or not the attributes are actually exemplified by something. Again, this was clear to medieval philosophers who said that *talia sunt subjecta qualia permittuntur a praedicatis suis* (i.e., individual entities are such as they are permitted to be by their attributes); but this is also the means for making existence claims in the sciences (and not only in the sciences). Indeed, if we come to the conclusion that such and such an entity should exist, we try to check, by appropriate criteria varying from science to science, whether certain sets of attributes are actually exemplified; and if this is not the case we cannot affirm the existence of the supposed entity.

The above considerations indicate that providing a list of simples does not amount to "naming" a set of unconnoted individuals, but to offering a list of *primitive predicates* corresponding to certain *basic attributes* of reality which we intend to consider as the nonanalysed starting point of our discourse. These attributes are contextually interrelated and (in this sense) interdefined; and the "referential simples" of our discourse are those entities that exemplify our set of basic attributes.

This, in particular, is the case in the sciences, and is the correct way of making precise what the *objects* of a particular science are. In this way we can see, first, that the notion of simplicity is not absolute but relative, depending on the particular choice of the primitive predicates adopted in a science, so that what is simple "from a particular point of view" (that is, from the point of view of a given science that has selected certain attributes as its objects of study and certain predicates for denoting them) may not be simple from another point of view, or within another science. And second, we see that the simples can be described, since they are only the structured set of particular attributes that are expressed linguistically by a corresponding structured set of predicates. When a science has set out to investigate reality from a certain point of view (that is, limiting its interest to a given set of attributes), several *things* of the world can be approached; and usually they can also be investigated, i.e. understood and explained as *complex*, in the sense that their properties can be shown to derive from the properties of the admitted simples, thanks to the *relations* constituting the structure of each complex thing.

2. Analytic and synthetic simplicity

We have said above that the notion of absolute simplicity is paradoxical and of little use. We must recognise, however, that it can play a significant role in characterising an important ontological notion, that of *individual*. Traditional ontology used to maintain that whatever exists has an identity and unity in itself (*indivisum in se et divisum a quolibet alio*). This does not prevent an individual entity's having parts that may be *distinguished*; these parts, however, cannot be *separated* without that entity ceasing to be what it is. This amounts to saying that any individual must be conceived of as a *whole* and, in this sense, it is ineffable because no description could account for the unlimited number of its properties and for their being brought into unity in that particular and unique way. For this reason we can say that not only "elementary entities" are simple, but also "wholes" are such. On the other hand, we consider wholes as typical examples of complex entities, and this creates a puzzle.

Help in solving this puzzle is afforded by the distinction between analytic and synthetic simplicity[1]. The intuitive idea that complexity entails relations among internal parts produces by opposition the notion of *analytic simplicity*: something is analytically simple if it has no *internal relations* (therefore, something is analytically complex if it *has* internal relations). If we recall what we have said about *wholes* that can be considered as simples, we must say that their simplicity cannot consist in not having internal relations (indeed they have), but rather in their not having *external* relations. Therefore we can define the notion of *synthetic simplicity* by saying that something is synthetically simple if it has no *external relations* (therefore, something is synthetically complex if it *has* external relations).

The usefulness of this distinction comes from the fact that analytic and synthetic simplicity are not absolute, but *relative*, since they depend on the point of view adopted: something can be (and usually is) both simple and complex due to the fact that we can consider it as having or not having

---

[1] As developed by Dilworth C.: 2001, "Simplicity", *Epistemologia,* 24/2, pp. 173-201.

internal or external relations. For example, atoms in traditional chemistry or in the ideal gas model are analytically simple (they have no internal parts or relations), whereas they are analytically complex in atomic physics (which investigates their internal structure). Nevertheless the conceptual fruitfulness or even the pure and simple rationale for introducing these analytically simple entities into science resides in the fact that they have been thought of as at the same time being synthetically complex, that is, as having external relations at least among themselves, such that the admission of their existence and their external relations could account for the properties of the "whole" process of which they are considered to be constituents.

3. The nature of complexity

This double face is essential for the understanding of complexity and, in particular, for understanding in what precise sense it is the contrary of simplicity. As a matter of fact, we have said at the beginning that 'complex' is the contrary of 'simple', but this is not completely true: if we consider the common use of language we find that 'simple' is often understood as the contrary of 'compound', which is not synonymous with 'complex'. The difference between these two concepts can be expressed by saying that in a *compound* we have a plurality of components, but are not concerned about their relations, whereas a *complex* is a compound in which the relations among its constituents are significant, since they make of this compound a *whole* endowed with an identity and evincing an analytical complexity. Said differently: an entity is genuinely *complex* if it contains components that are analytically simple but at the same time synthetically complex in such a way that their external relations coincide (at least in part) with the internal relations in which the analytical complexity of the whole consists.

This interplay of the conditions of analytic simplicity and synthetic complexity (and analytic complexity and synthetic simplicity) allows us to understand that the idea of an *order*, of a *structure*, is intrinsic to the notion of complexity: we are prepared to say that, *from a particular point of view*, a building is complex while a heap of bricks is not (but is, rather, *chaotic*); and this not so much because the bricks in the heap have no relations (they have some spatial relations, for example), but because, from this point of view, these relations do not make up an ordered structure. They do not correspond to the "internal relations" of the heap (which *is* a heap precisely because does not entail any particular structure). This difference is considered, for example, in chemistry, where mixtures are distinguished from combinations: in mixtures several components are present in different proportions, but they maintain their specific properties, and the "whole" does not evince properties of its own that are not possessed by its parts. In a combination, on the other hand, the components are brought to a unity via a chemical reaction, and the result is a substance whose properties are new and even very different from the properties of its components. Both mixtures and combinations are chemical *compounds*, but only the results of combinations can be said to be complex.

From the above it follows that the notion of complexity applies in a proper way to realities that are taken as *wholes* and are, at least to some extent, considered as synthetically simple, since they show specific properties as wholes, properties that are expressed by means of particular primitive predicates of the language intended to denote certain basic attributes. Such wholes, however, must be characterised or characterisable through certain *internal relations* that constitute an order, such that they can be viewed as being analytically complex as well. The ideal goal of this study is that the global properties of the whole can be shown to be the effect of the properties of the analytically simple (but synthetically complex) parts. If this goal is attained we can speak of a successful *reduction*, in the sense that the properties of the simples are *sufficient* (on the basis of the recognised relations subsisting among these simples) for producing (ontologically) and explaining (epistemologically) the properties of the complex. An example of this sort is traditional celestial mechanics, where the laws and properties of the solar system were shown to be a consequence of the physical laws of Newtonian mechanics, understood as the expression of the relations among certain "simples" (the mass-points) that were "exemplified" by the celestial bodies. In this case the reduction was not surprising, since no new attributes were used for singling out the "whole" (the solar system), which was considered only from the point of view of attributes such as mass, position in space and time, motion, and gravitational force.

4. Reduction and emergence

More interesting is the case where the attributes characterising the whole are not the same as those characterising its analytic simples. In the kinetic theory of gases, for example, attributes such as volume, pressure and temperature that characterise the state of a gas in a vessel, are explained as the effect of the chaotic motion of the gas molecules, of their impacts on the vessel's walls, of the mean kinetic energy of the molecules that themselves, however, have no volume, pressure, or temperature. It is debatable whether this reduction of the properties of the whole to those of the parts, in this specific case, is fully successful. Let us suppose that it is: in this case we would have an interesting instance of reduction where the study of a possible model of analytic complexity for a whole has the consequence of eliminating a *limit*: those properties that initially were taken as characteristic of the whole and "primitive" (i.e., indefinable) continue to remain characteristic of the whole, but appear as consequences of the properties of the simples, thanks to the analytic complexity of the whole.

It seems to me that in this case it would be appropriate to speak of *emergence* in a convincing sense, since these properties "emerge" from those of the simples in an *understandable* way. The customary use of "emergence", however, is different: it is used to indicate the presence of properties that can *not* be explained as the consequence of the properties of the analytic simples. One is better advised not to abandon established conventions, unless for serious reasons. Therefore we will adhere to this notion of emergence and speak of *resultance* in the case of properties of the whole that are produced by properties of the analytic simples by virtue of certain internal relations of the whole. As a consequence, emergence appears as the contrary of reduction; but it must be noted that emergence is not, by itself, a corollary of complexity: complex entities sometimes show properties that are reducible to those of their analytic simples (resultant properties), and sometimes do not (emergent properties).

As a general characterisation of our discussion we can say that a complexity approach always presupposes two "limits" that are constituted by simples. One is the "superior limit" of the whole that, as we have said, functions as a simple in that, while it is characterised by certain attributes, it is not considered with regard to its external relations (i.e. it is *synthetically* simple). The other is the "inferior limit" of the simples that are admitted in that approach: they are *analytically* simple and also characterised through certain attributes. The complexity approach consists, as we have explained, in bridging these two limits by a thought process in which the analytically simple components (with their "external" properties) can be treated as synthetically complex, so that their external relations make up the internal relations of the whole when it is considered as analytically complex from that particular point of view.

This general framework can be used in different ways according to one's point of view, i.e. according to the choice of the two poles or simples introduced: if the attributes of the whole and the components are the same (or at least belong to the same "family") we have the situation common in most sciences in which a complex reality is studied, interpreted, and explained by "analysing" it into its constituent parts and recognising "synthetically" how the properties and functions of these parts contribute to the properties and functions of the whole (the simplest example is perhaps that of the study of an organism in biology, where its simples are taken to be its organs or even its cells, that are the "minimal units" to which the attribute of life can be assigned). In other cases the attributes characterising the whole are not of the same kind or family as those characterising its proposed analytic simples (for example, when the analytic simples of a living organism are considered not to be its cells but its molecules or even atoms). Here the challenge is much greater, for the superposition of the patterns of the analytic complexity of the whole and of the synthetic complexity of the simples may be hard to show. Indeed it is usually easy to "analyse" a whole into ever more elementary parts, and see *how* they cooperate according to the *given* internal structure of the whole (this is essentially a descriptive task), but much more difficult to show *why* the particular analytic simples should or could give rise to certain special relations, and how this fact should lead to new, internally complex, synthetic simples endowed with emergent properties, in an ascending hierarchy of levels terminating in the constitution of the highest whole with its special properties.

The best we can expect from the descending analysis of this whole is that its analytic simples (with their "external" properties) are *necessary* to the existence of the structured whole; but we usually cannot show that they are also *sufficient*: first because we should show that the *structure* of the whole follows from the properties of the analytic simples, and second because it is by no means certain or evident that the appearance of new attributes is just a question of structure. Again, the complexity of the structure is *necessary* for the appearance of the new attributes, but it must be proved that it is also sufficient.

We have reached in this way what seems to be our most prudent and critical conclusion regarding the relation between complexity and emergence: complexity is certainly a *necessary* condition for emergence; and in attempting to understand the novelties of the attributes possessed by certain entities in comparison with other less complex entities the study of their complexity is of great help. But this does not amount either to *explaining* these novelties in terms of the attributes of the less complex (analytically simple) entities, nor even to *explaining away* such novelties. In other words, variety of attributes is what characterises reality and makes it interesting and beautiful, and one sees no point in attempting to eliminate this variety. Complexity is a fruitful path for understanding how this variety is composed in a *unity*, but this is different from *uniqueness*. Several different *units* are possible, each characterised by its special attributes as a whole, that grant it a synthetic simplicity, and endow it with internal and external relations linking it with other units characterised by other attributes, in a complex net in which emergent attributes are not dissolved but harmonised.

Reference

Dilworth, C.: 2001, "Simplicity", *Epistemologia,* 24/2.

2. On Levels and Types of Complexity and Emergence

Hans Lenk and Achim Stephan

Dept. of Philosophy, University of Karlsruhe, Germany

Our paper on complexity and emergence is primarily epistemological and methodological, epitomising some epistemological perspectives on current approaches in scientific research on dynamic complex systems, notably so-called "self-structured" or "self-organized" systems that manifest emergent properties, structures and processes, which are very common in natural as well as social contexts. We will not be studying complexity and emergence from a scientific point of view in the narrow sense by providing new scientific or substantial insights into ontological, micro- or macro-approaches and structures, nor into quantum interpretations. Rather, some epistemological perspectives will be outlined which may shed new light on the traditional debate regarding determinism, probability, emergence and complexity from a methodological point of view. Some new insights from deterministic chaos theory, as well as insights regarding problems of treating reductionism and supervenience, will be used, but not elaborated in detail. The overall insight may be summarized as consisting in the realization of the methodological necessity of distinguishing levels of complexity, interpretations and approaches, and contexts in which parts and substructures are embedded in complex structures and their respective relativizations. The insight also includes, and most importantly, the appreciation of the need for epistemological sophistication or indirectness with regard to the methodology of procedural approaches and theoretical concepts. The emphasis is thus critical, criticizing naive ontology as well as direct realistic projections of structures to be theoretically described and interpreted into or onto the world as such.

In our approach, however, we are realists and naturalists of a sort, though not reductive physicalists in a simple sense, but rather methodologically sophisticated quasi-Kantians, though admitting and postulating a kind of modification and liberalization of Kant's to our mind too rationalistic conception of *the* structure of reason in an *a priori* prefabricated and/or logical provenance. We shall in our rather sketchy presentation address problems of systems analysis and complexity studies, as well as emergence and questions related to various methodological approaches and perspectives on interpretation and the introduction of different levels and concepts of complexity, etc. Thus, the approach will be a rather perspectivistic or, methodologically speaking, structural-interpretationist one, relativised by the pertinent theoretical and experimental setup used. For example, there are no interesting (complex) deterministic systems *per se,* but primarily systems which more or less allow for a deterministic, theoretical approach. Thus the questions concerning determinism versus probabilism, micro- and macro-analysis, and chaotic phenomena are all dependent on epistemological presuppositions and methodological higher-level interpretations. (Only general negative judgments can be definitively put forward, for instance that most interesting systems in the world may not be amenable to a deterministic analysis).

1. Complexities on different (meta-)levels

The world and most subsystems in the universe are complex and show some kind of non-aggregateness if there are at least three or more degrees of freedom involved, i.e., in physicists' terms, more than two dimensions of the correlated state spaces. Indeed, traditional ontology maintained that there are different levels of structures of beings (or complex systems). This is, from a naive point of view, certainly true; but it must be refined with respect to susceptibility to analysis and theoretical approaches as well as to levels of complexity. Not only are different concepts of complexity available from a mathematical (algorithmic), computational, physical and empirical as well as a chaos-theoretical point of view (see e.g. Leiber 1998, 1999, 2001), but we also have epistemologically and methodologically to take into account different levels of meta-theoretical structuring and steps or levels of theoretical interpretations, meta-interpretations etc. Mathematical, computational (in principle or practice), as well as physical limitations may have an important influence on the availability of and susceptibility to analysis proper of the problems of complex systems, but from a methodological point of view these are basically still to be located on one or two levels. The meta-levels mentioned will each in turn also give rise to problems of complexity. Therefore we have to distinguish between problems of the different types of complexity on the basic object-language level of theories on the one hand, and methodologically induced or impregnated complexities of a meta-theoretical or higher level on the other.

Generally speaking, the approach taken by general complex systems theory and, e.g., deterministic chaos theory are, epistemologically speaking, certainly very interesting with respect to problems of reductive explanations, new sorts of simulation-explanations, predictions (or rather the unpredictability of single states in phases or state spaces of chaotic character, i.e. admitting of strange attractors). Epistemologically speaking we have a limitation of scientific expectations for long-range predictions of individual trajectories of systems, and the respective in-principle or practical limitation of the treatability of computation and problem solving. Clearly these new systems disciplines, including chaos theory, do not give rise to fundamentally new basic theoretical approaches or revolutions, such as the one created by quantum theory in the last century; but epistemologically speaking—and thus for philosophy of science—they will have rather substantial impact with respect to the long-range treatment of physical, biological and social systems, as well as with respect to epistemology and philosophy of science in general (Kanitscheider 1995). No new model of physics or natural science is involved, but a new level of discussion for the interpretation and metatheoretical analysis of the sciences, and for their structural models and theoretical as well as methodological questions. There might even be challenges to develop new theoretical models to approach certain systems. Thus far, chaos theory and fractal geometry are relevant only to deterministic systems (or systems susceptible of a deterministic analysis), while many systems in the world—e. g. social and economic systems—are rather probabilistic or Markovian, with some additional constraints. To date, we have no chaos theory for probabilistic systems, though they certainly show "chaotic" phenomena too, even more so than so-called deterministic systems.

Excursion into the question of levels of structure and schema interpretations for metatheoretical complexity studies

Before concentrating on the specific topic of emergence, let us give a short overview of the basic approach of methodological structural levels, or schema interpretationism, as developed by one of the authors during the last few decades. These perspectives will, epistemologically speaking, be relevant for subsequent discussions concerning philosophy of science approaches to complex dynamical systems, including phenomena of emergence, chaotic phases, etc.

Interestingly enough, methodological questions and approaches of structural models or schema-interpretations would overarch even the modern split between the natural and social sciences as well the humanities, since all these disciplines would structure their fields and objects according to the activation of schemata or structural models by using procedures to establish, stabilize, and activate schemata and models as cognitive constructs in order to structure the respective world versions and sets of objects events, and structures, as well as procedures and projections.

It is interesting that schema interpretation admits of levels of categorization according to the variability of the respective schemata, i.e. whether or not they are hereditarily fixed or conventionalized or flexible, whether they are subconsciously developed and activated or consciously conceived and used. One of the present authors developed a hierarchy of levels of interpretation consisting of six different levels or planes of interpretation. The following diagram shows the six[1] levels:

Diagram of the levels of interpretation

IS$_1$: practically unchangeable *productive primary interpretation* (*"Urinterpretation"*) (primary constitution or schematization, respectively)

IS$_2$: habit-shaping, (equal) form-constituting *pattern interpretation* (ontogenetically habitual(ized) form and schema categori(ali)zation and preverbal concept-formation)

IS$_3$: conventional concept formation transmitted by social, cultural and norm-regulated tradition

---

[1] In mutual correspondence (as of Jan. 12, 1987) and cooperation Abel parallely developed only three levels: $I_1 = IS_1$; $I_2 = IS_2 \cup IS_3$; $I_3 = IS_4 \cup IS_5 \cup IS_6$ (see e. g. Abel 1993).

IS$_{3a}$: ... by *non-verbal* cultural gestures, rules, norms, forms, conventions, implicit communicative symbols

IS$_{3b}$: ... by *verbal* forms, and explicitly representing communicative symbols, metasymbols, metaschemata etc.

IS$_4$: applied, *consciously* shaped and accepted as well as transmitted *classifactory interpretation* (classification, subsumption, description by "sortals", generic formation of kinds, directed concept-formation)

IS$_5$: explanatory and in the narrow sense "comprehending" ("*verstehende*"), *justifying, theoretically* or *argumentatively substantiating interpretation*; *justificatory interpretation*

IS$_6$: epistemological (methodological) *metainterpretation* (plus meta-metainterpretation etc.) of methods, results, instruments; the conception of establishing and analyzing interpretive constructs themselves

The different levels of interpretation are the following: IS$_1$ comprises the practically unchangeable productive primary interpretations of primary constitution which might be represented by subconscious schema instantiation. They constitute the hereditarily fixed or genetically founded activation of selective schemata of sense perception (e. g. contrasts of dark and light etc.) as well as the interactive, selective activations of early ontogenetic developments such as the stages of developmental psychology discussed by Piaget. Also comprised are the biologically hardwired primary theories which we cannot alter at will, but which we can (only) problematize in principle. For instance, we have no magnetic sense or capacity to trace ultrasound as do bats. But we can conceive of conditions in which we *could* have such senses, or at least devise technological means for substituting for these.

On the second level we have the habitual, quality-forming frame interpretations and schema categorizations, as well as categorializations which are abstracted from prelinguistic discriminatory activities, experiences of equality of shape, similarity of presentation and experience, etc. The establishment and discriminatory capacity of prelinguistic conceptualization and development of concepts about language is to be formed on this level.

On level IS$_3$ we have conventional concept formation, i.e. traditional social and cultural conventions and norms for representation, as well as forms of discriminatory activities such as the explicit conceptualization of framing the world according to natural kinds etc. In so far as this is not already related to language differentiation we can think of a sublevel (IS_{3a}) on which prelinguistic conventionalizations are characteristic. On the other hand (on IS_{3b}) we have the explicitly linguistic conventionalization or the differentiation of concepts by means of language.

Level IS_4 would comprise the consciously formed interpretations of embedding and subsuming as well as classifying and describing according to generic terms, kinds, etc. This is the level of ordered concept formation and classification as well as ordering and subsumption.

Level IS_5 would go beyond this by rendering explanatory, or in the narrower sense comprehending (*"Verstehen"*) interpretations, as well as justifying and theoretically arguing for interpretations, in the sense of looking for reasons and grounds of justification.

These activities are certainly not only advanced in science and intellectual disciplines but also in everyday life and common sense. Any kind of systematic comprehension within the area of theories, systems and overarching perspectives of integration is important here.

Beyond this, however, we also have a level IS_6 of the epistemological and philosophical as well as methodological interpretations of a meta-character, overarching and integrating the procedures of theory-building and theory interpretation, methodology and the models of interpretation in the sense of methodological schema interpretationism itself. One could call this a metalevel of interpretativity and speak of epistemological metainterpretations. However, this level is cumulative and can be considered as being open toward further meta-levels. The model and approach of epistemological interpretationism is itself certainly an interpretive one and can be described and developed only on a certain meta-level which is to be seen within the level IS_6. Therefore, we have the possibility of a self-application of the interpretational method to interpretive procedures themselves. The philosophy of schema interpretation is a philosophy of interpretational constructs as an epistemological model which admits of a certain kind of metatheoretical and metasemantical self-application in the form of a sort of "metainterpretation".

Structural model and schema interpretation is not everything, but anything conceivable is perspectivally interpretation-dependent, or in the more specific sense interpretation-laden, if not even—as in the case of direct perception—schema interpretation-impregnated in the narrower sense. Everything can only be grasped by means of schema interpretation, i.e. by constituting schemas and developing as well as activating and reactivating schemas. Any "grasping" of anything whatsoever (be it seemingly passively in the form of perception or "impregnation" in the narrower sense by factors of the 'external world', or more actively by framing thoughts and actions) is formed, influenced or externally impregnated by schema selection and activation. This approach is particularly relevant to the neurosciences, philosophy of mind, and any other emergent phenomena in complex systems of self-organization or self-structuredness to be found in natural and social setups.

Different versions of emergentism and problems of reduction, notably for the philosophy of mind

As W. Wimsatt 1996 has repeatedly stressed, "emergence—as a failure of aggregativity—is extremely common". Most systems in the world admitting of a state phase description of more than two dimensions (or with three or more degrees of freedom) do not avail themselves of a linear decomposition in terms of the aggregation of only simple components. Most systems are non-aggregative; lacking "aggregativity", therefore, by definition, they admit of emergent phenomena, properties, structures, and processes. Wimsatt characterizes "aggregativity" in terms of four fundamental conditions: 1. Rearrangement or intersubstitution of parts, 2. Independence of size scaling, 3. Invariance under the decomposition and reaggregation of parts, and 4. Linearity (no mutual interactions among the parts for the production or realization of the system property).

According to Wimsatt, emergence is defined by the failure to meet all of these conditions[2]. However, he does not pay attention to different sorts of emergentism and their respective epistemological problems. This will be the main topic of the second part of this paper.

---

[2] By the way, as Wimsatt 1996 himself admits, conditions 1 and 3 might be combined to invariance of the system's property under interchanging any number of parts and under decomposition and reaggregation. In short, we can say that the composition function of aggregative systems should be commutative, associative and additive (linear) as well as recursively generalizable.

2. A whole bunch of emergentisms

While in ordinary language 'to emerge' means 'to appear' or 'to come into view', in most technical uses the term 'emergent' denotes a second-order property of certain first-order properties (or structures), namely those that are emergent.

However, what the criteria are by which emergent phenomena should be distinguished from non-emergent phenomena is quite controversial. Among these features are novelty, unpredictability, irreducibility, and the unintended arising of systemic properties, particularly in artificial systems. Some of these criteria are very strong, so that few, if any, properties count as emergent. Other criteria are inflationary in that they count many, if not all, system properties as emergent. One of the consequences of this controversy is a great confusion about what is really meant by an 'emergent property', when this term is used in such different disciplines as philosophy of mind, theories of self-organization, dynamical systems theory, artificial life, and connectionism.

Varieties of emergence

There are three theories among the different varieties of emergentism deserving particular attention: *synchronic* emergentism, *diachronic* emergentism, and a *weak* version of emergentism. For synchronic emergentism the relationship between a property of the system and its microstructure, i.e. the arrangement and the properties of the system's parts, is at the center of interest. For such a theory, a property of a system is taken to be emergent if it is *irreducible*, that is to say, if it is not reductively explainable. In contrast, diachronic emergentism is mainly interested in the *predictability* of novel properties. For such a theory, those properties are emergent which could not, in principle, have been predicted before their first instantiation. Both stronger versions of emergentism are based on a common weak theory from which they can be developed by adding further claims.

Weak emergentism

The first feature of contemporary theories of emergence—the thesis of *physical monism*—is a thesis about the nature of systems that have emergent properties or structures. The thesis says that the bearers of emergent features consist of material parts only. Thus, all substance-dualistic positions are rejected, for they base properties such as being alive or having cognitive states on supernatural bearers such as an *entelechy* or a *res cogitans*, respectively.

> (PM) *Physical monism*. Systems that exist or come into being in the universe consist solely of material entities. Likewise, properties, dispositions, behaviors, or structures classified as emergent are instantiated by systems consisting exclusively of physical entities.

While the first thesis puts the discussion of emergent properties and structures within the framework of a physicalistic naturalism, the second— the thesis of *systemic properties*—delimits the type of properties that are possible candidates for emergents. It is based on the assumption that *general* properties of complex systems fall into two different classes: (i) properties which some of the system's parts also have, and (ii) properties that none of the system's parts have. These properties are called systemic or collective properties.

> (SP) *Systemic properties*. Emergent properties are systemic properties. A property is systemic if and only if a system possesses it, but no part of the system possesses it.

It should be uncontroversial that both artificial and natural systems with systemic properties exist. Those who would deny their existence would have to claim that *all* of a system's properties are already instantiated by some of the system's parts. Countless examples refute such a claim. The third thesis specifies the type of relationship that holds between a system's micro-structure and its emergent properties as a relationship of *synchronic determination*:

> (SD) *Synchronic determination*. A system's properties and dispositions depend nomologically on its micro-structure, that is to say, on its parts' properties and their arrangement. There can be no difference in the systemic properties without there being some differences in the properties of the system's parts or in their arrangement.

Anyone who denies the thesis of *synchronic determination* either has to admit properties that are not bound to the properties and arrangement of its bearer's parts, or she has to suppose that some other factors, in this case non-natural factors, are responsible for the different dispositions of systems that are identical in their micro-structure. She would have to admit, for example, that there may exist objects that have the same parts in the same arrangement as diamonds, but which lack diamonds' hardness. This seems to be implausible. Equally beyond thought is that there may exist two micro-identical organisms, one of which is alive and the other not. In the case of mental phenomena, opinions may be more controversial (think of externalism concerning the content of mental states); but one thing seems to be clear: anyone who believes, e.g., that two creatures identical in micro-structure could be such that one is colorblind while the other is not, does not hold a physicalistic position.

Weak emergentism as sketched so far comprises the minimal conditions for emergent properties. It is the common base for all stronger theories of emergence. Moreover—and this is a reason for distinguishing it as a theory in its own right—it is held not only by some philosophers such as Mario Bunge and Gerhard Vollmer, but precisely in its weak form also by cognitive scientists such as John Hopfield, Eleanor Rosch, Francisco Varela, and David Rumelhart. Weak emergentism, however, is compatible with contemporary reductionist approaches without further ado. Some champions of the *weak* as versus *stronger* versions of emergentism credit the compatibility of emergence and reducibility as one of its merits.

Synchronic emergentism

The essential features of more ambitious theories of emergence are those of *irreducibility* and *unpredictability*, respectively. Both are closely connected: irreducible systemic properties are *eo ipso* unpredictable before their first appearance. Hence, synchronically emergent properties are also diachronically emergent, but not *vice versa*.

A systemic property is irreducible if it cannot be explained reductively. For a reductive explanation to be successful, several conditions must be met: (a) the property to be reduced must be functionally construable or reconstruable; (b) it must be shown that the specified functional role is filled by the system's parts and their mutual interactions; and (c) the behavior of the system's parts must follow from the behavior they show in isolation or in simpler systems than the system in question. If all conditions are fulfilled, the behavior of the system's parts in other contexts reveals what systemic properties the actual system has. Since these conditions are independent, different possibilities for the occurrence of *irreducible* systemic properties result:

(I) *Irreducibility*. A systemic property is irreducible if (a) it is not functionally construable (or reconstruable), or if (b) it cannot be shown that the interactions among the system's parts fill the systemic property's (re)construed functional role, or if (c) the specific behavior of the system's components, over which the systemic property supervenes, does not follow from the component's behavior in isolation or in simpler constellations.

Thus, we have to distinguish three different types of irreducibility of systemic properties. Equally different seem to be the consequences that result from them. If a property is irreducible due to the irreducibility of the behavior of its bearer's parts, we seem to have an instance of 'downward causation'. For, if the components' behavior is not reducible to their arrangement and the behavior they show in simpler systems, then there seems to exist some 'downward' causal influence from the system itself or from its specific structure on the behavior of its parts. However, if there should exist such instances of 'downward causation' this would not amount to a violation of some widely held assumptions, such as, for example, the principle of the causal closure of the physical domain. Within the physical domain, we would just have to accept additional types of causal influence besides the known types of mutual interactions. Likewise, if it cannot be shown that the interactions between the system's parts fill the specified functional role, it seems that the systemic property has causal powers the microstructure lacks; hence, there would be some downward causal influence as well.

In contrast, the occurrence of properties that are not functionally construable does not imply any kind of downward causation. Systems that have functionally unanalyzable properties need not be constituted in a way that amounts to the irreducibility of their components' behavior. Nor is it implied that the system's structure has a downward causal influence on the system's parts. All the more, there is no reason to assume that unanalyzable properties themselves exert a causal influence on the system's parts. Rather it is to ask how functionally unanalyzable properties might have any causal role to play at all. And, if one can not see *how* they might play a causal role, then, it seems, such properties are epiphenomena.

Diachronic emergentism

All diachronic theories of emergence have at bottom a thesis about the occurrence of genuine *novelties* in evolution. All preformationist positions are excluded.

> (N) *Novelty*. In the course of evolution, exemplifications of genuine novelties occur again and again. Already existing building blocks develop new constellations; new structures are formed that constitute new entities with new properties and behaviors.

However, the bare addition of the thesis of novelty does not turn a weak theory of emergence into a strong one, since reductive physicalism remains compatible with such a variant of emergentism. Only the addition of the thesis of *unpredictability in principle*, will lead to stronger forms of *diachronic* emergentism.

The first occurrence of a systemic property can be unpredictable for different reasons: (a) it is unpredictable, in principle, if it is irreducible. This does not preclude, however, that further occurrences of the property might be predicted adequately; (b) it can be unpredictable because the micro-structure of the system that exemplifies the property for the first time in evolution is unpredictable. Since in the first case criteria for being unpredictable are identical with those for being irreducible, this notion of unpredictability will offer no theoretical gains beyond those afforded by the notion of irreducibility. Let us focus, therefore, on the second case: *unpredictability of structure*.

The structure of a newly formed system can also be unpredictable for two reasons. If the universe is indeterministic, then its novel structures will *eo ipso* be unpredictable. However, from an emergentist perspective it would be of no interest if a new structure's appearance were unpredictable only as a result of its indeterminacy, not to mention that most emergentists claim the development of new structures to be governed by deterministic laws. Never-theless, deterministic formings of new structures can be *unpredictable in principle* if they are governed by laws of deterministic chaos. Against this view one might argue that a Laplacean demon could correctly predict even chaotic processes. Whether or not this 'actually' could be the case is not yet settled. It depends mainly on the question of what kind of information we allow such a creature of fantasy to have. We can at least preclude that foretellers of our mental capacities have these abilities; and thus, we can suppose that where chaos exists, structures exist that are unpredictable in principle.

(SU) *Structure unpredictability*. The rise of a novel structure is unpredictable in principle, if its formation is governed by laws of deterministic chaos. Likewise, any property that is instantiated by the novel structure is unpredictable in principle.

Although structure emergentism implies the unpredictability of all properties instantiated by systems that emerge from chaotic processes, it does not thereby imply their irreducibility. As far as that goes, the unpredictability, in principle, of systemic properties is entirely compatible with their being reducible to the micro-structure of the system that instantiates them.

Synopsis

The following figure depicts the logical relationships that hold among the different versions of emergentism.

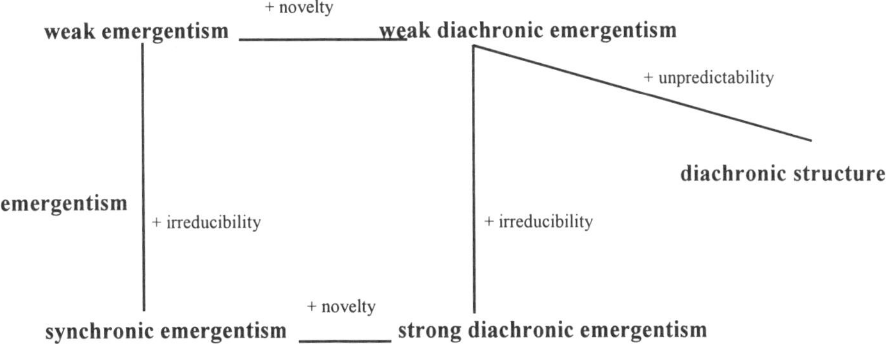

Weak diachronic emergentism results from *weak emergentism* by adding a temporal dimension in the form of the thesis of novelty. Both versions are compatible with reductive physicalism. Weak theories of emergence are used today mainly in the cognitive sciences, particularly for the characterization of systemic properties of connectionist nets, and in theories of self-organization. *Synchronic emergentism* results from weak emergentism by adding the thesis of irreducibility. This version of emergentism is important for the philosophy of mind, particularly for debating nonreductive physicalism and qualia. It is no longer compatible with reductive physicalism. *Strong diachronic emergentism* only differs from synchronic emergentism because of the temporal dimension in the thesis of novelty. In contrast, *structure emergentism* is entirely independent of synchronic emergentism. It results from weak emergentism by adding the thesis of structure unpredictability. Although structure emergentism emphasizes the boundaries of prediction within physicalistic approaches, it is compatible with reductive physicalism, and so it is weaker than synchronic emergentism. Theories of deterministic chaos can be acknowledged as a type of structure emergentism. Likewise, its perspective is important for evolutionary research.

Arguments pro and con emergence

We take it to be clear that weak emergent properties exist. At best, one might ask why such properties should be called 'emergent' at all, and not just 'systemic'. Furthermore, since chaotic processes of structure formation exist, structure emergence also exists. Thus, what we really need is an argument for synchronic emergence, i. e. an argument for the existence of properties that are not and that will not be reductively explainable. Many natural scientists deny the existence of such properties, since they do not know of any properties that could not be reductively explained, at least in principle. All systemic properties studied in the natural sciences are, without exception, functionally construable, their functional roles always being filled by the interactions of their systems' parts, and the behavior of the parts of any system always seems to follow from their behavior in simpler systems. Therefore, some critics question whether it is plausible to develop the notion of synchronic emergence in the first place. But, even if it should turn out that all systemic properties studied in the natural sciences are reductively explainable, it is useful to have the strong notion of synchronic emergence. More than any other notion it is apt to clearly formulate nonreductive positions concerning the mind-body problem.

Whether or not synchronically emergent properties actually exist does not seem to depend on empirical, but rather on conceptual grounds. Among others, Broad 1925, Levine 1993, and Chalmers 1996 have argued forcefully that properties such as qualia are not functionally analyzable. Thus, to the extent that their arguments are convincing we have good reasons to assume that phenomenal qualities are emergent properties in the strong sense.

However, if mental properties such as qualia are emergent in the strong sense, then new problems arise. Some philosophers have claimed that irreducible properties *eo ipso* exert downward causation. In the case of mental phenomena, however, this would conflict with the principle of the causal closure of the physical domain. But, as we have seen above, properties that are irreducible for conceptual reasons do not imply downward causation. Rather, they give rise to another objection: how can properties that escape reconstruction via their causal role play any causal role at all? The reply to this objection depends mainly on what concept of causation we have. If we think supervenient causation suffices for causation, then irreducible emergent properties can be causally efficacious. If we think supervenient causation does not suffice, then irreducible emergent properties do not seem capable of playing any causal role. But these questions are not yet settled (see Stephan 1997).

3. Looking back and ahead

We gave an outline of both the complexity we encounter in the world and the complexity of the descriptions and interpretations we develop to make sense of that complex world. Furthermore, we explored and discussed a variety of different notions of emergence. These notions perform rather different tasks within the philosophy of mind, the philosophy of science and different fields of cognitive science. Thus, philosophers have a lot of work before them to investigate and carefully interpret what colleagues do with the tool of emergence in their specific subjects.

References

Broad, Ch. D.: 1925, *The Mind and Its Place in Nature*. London: Kegan Paul.
Chalmers, D. J.: 1996, *The Conscious Mind*. Oxford: Oxford University Press.

Kanitscheider, B.: 1995, "Die Relevanz der Chaos-Theorien für die Philosophie", in Lenk, H.and Poser, H. (eds.): 1995, *Neue Realitäten-Herausforderung der Philosophie*. Berlin: Akademie, pp. 169-184.

Kim, J.: 1999, "Making Sense of Emergence", *Philosophical Studies* 95, pp. 3-36.

Leiber, T.: 1998, "On the Actual Impact of Deterministic Chaos", *Synthese* 113, pp. 357-379.

— 1999, "Deterministic Chaos and Computational Complexity: The case of methodological complexity reductions", *Journal for General Philosophy of Science* 30, pp. 87-100.

— 1999,2001 "On the Impact of Deterministic Chaos on Modern Science and Philosophy of Science: Implications for the Philosophy of Technology?", in Agazzi, E. and Lenk, H. (eds.): 1999, *Advances in the Philosophy of Technology*. Newark, DE, University of Delaware/Society for Philosophy and Technology, pp. 57-186, and also in Lenk, H. and Maring, M. (eds.): 2001, *Advances and Problems in the Philosophy of Technology*. Münster: LIT, pp. 141-166

Lenk, H.: 1993, *Interpretationskonstrukte: Zur Kritik der interpretatorischen Vernunft*. Frankfurt a. M.: Suhrkamp.

— 1995, *Interpretation und Realität*. Frankfurt a. M.: Suhrkamp.

— 1995a, *Schemaspiele. Über Schemainterpretationen und Interpretationskonstrukte*. Frankfurt a. M.: Suhrkamp.

— 1998, *Einführung in die Erkenntnistheorie: Interpretation-Interaktion-Intervention*. München: UTB (Fink).

— 2000, *Erfassung der Wirklichkeit. Eine interpretationsrealistische Erkenntnistheorie*. Würzburg: Königshausen u. Neumann.

Levine, J.: 1993, "On Leaving Out What It's Like", in M. Davies and G. W. Humphreys, (eds.): 1993, *Consciousness*. Oxford: Blackwell, pp. 121-136.

Stephan, A.: 1997, "Armchair Arguments Against Emergentism", *Erkenntnis* 46, pp. 305-314.

— 1998, "Varieties of Emergence in Artificial and Natural Systems", *Zeitschrift für Naturforschung* 53c, pp. 639-656.

— 1999, *Emergenz. Von der Unvorhersagbarkeit zur Selbstorganisation*. Dresden: Dresden University Press.

Wimsatt, W.C.: 1996, "Emergence as Non-aggregativity and the Biases of Reductionisms", in Taylor, P.J. and Haila, I. (eds.): 1996, *Natural contradictions*.

— 1997, "Aggregativity: Reductive heuristics for finding emergence", *Philosophy of Science* 64, Supplementary volume (PSA 96, Part II), pp. 372-384.

3. Formal Metatheoretical Criteria of Complexity and Emergence

C. Ulises Moulines

Dept. of Philosophy, University of Munich, Germany

One possible way to explicate complexity and emergence is to take a formal-ontological stance. I propose to view complexity and emergence as ontological categories applicable to empirical entities, either considered individually or else as members of kinds. I propose further to view them as *relational* notions. To abbreviate, let's symbolize both relations by "C" and "E", respectively. Then the fundamental logical form of complexity and emergence is this: $C(Y,X)$ (to be read as "Y is at least as complex as X"), and $E(Y,X)$ (to be read as "Y emerges from X").

What else can we say about the formal constitution of C and E? It seems to be an analytical truth, or at least a highly plausible explication of our intuitions, to settle C formally as a reflexive and transitive relation. It is certainly neither anti-symmetric nor connected, since different things may be equally complex, and there may be two different things which are not comparable with respect to their degree of complexity. (For example, it is very likely that it makes little sense to ask whether a painting is more or less complex than a symphony). So, complexity induces a *weak ordering* in the universe of empirical things. As for emergence, it appears to be a stronger ordering relation than complexity: it is asymmetric (if Y-things emerge from X-things, then X-things don't emerge from Y-things)[1], and it seems also plausible to characterize it as transitive: if Z-things are regarded as having emerged from Y-things and the latter as having emerged from X-things, then one might also say that, in the final analysis, Z-things have emerged from X-

---

[1] According to some theological or metaphysical systems, like those of Plotinus or Hegel, one might say that God or the Absolute Mind emerges from itself—but the reader may forgive me if I decide not to consider these sorts of transempirical entities here.

things. So, emergence induces a strict ordering in our universe. But, of course, the ordering is not total because the relation E is clearly not connected either: neither symphonies have emerged from paintings nor paintings from symphonies.

It seems to me that, at this point, we have already reached the maximum of general conceptual determination of the notions of complexity and emergence possible by purely conceptual analysis. But, of course, the formal scheme just laid out is still very poor in content, and, as general ontologists, we would like to find out more about complexity and emergence. In particular, we probably have the intuition that complexity and emergence are somehow related to each other. The intuition might go like this: emergence is a sufficient condition for greater complexity; or, viewed the other way around, greater complexity is necessary for emergence. Formally:

$$(\textit{EC}) \qquad\qquad \forall X \forall Y (E(Y,X) \rightarrow C(Y,X))$$

We could certainly add this as a further formal axiom to the conceptual determination of complexity and emergence, which would provide us with some information about the way these two notions are connected. Intuitively, there seem to be many examples to support this intuition: since atoms emerge from elementary particles, the first are more complex than the latter; since molecules emerge from atoms, they are more complex than the latter; living matter is more complex than inorganic matter because it has emerged from the latter, etc. However, it is not at all clear that it is a conceptual truth about complexity and emergence that this should always be the case. Unless we know more about both notions, why should we think that the concept of emergence already contains the concept of complexity?

I think the safest strategy for knowing more about complexity and emergence, and their possible mutual connections, is to analyze the scientific theories in which the two concepts seem to play a role, and see how they are applied. More concretely, we want to find out what it might mean for well-established empirical theories, that the domains of entities specific of one of them are more or less complex than those of another, and what it might mean that some of these domains emerge from others. We have then the following general metatheoretical question: What might complexity and emergence as inter-theoretical relations look like?

Now, a theory is something uniquely associated with a homogeneous class of models[2]. Here we are only interested in *empirical* theories, i.e. in theories such that at least some of their models can be regarded as more or less good representations of a given field of our experience. In a more technical terminology, this means that this field can be *subsumed* under at least one of the theory's models in such a way that some of the predictions made by means of the model really fit, at least approximately. The process of subsumption of a bit of experience under a theory's model is quite an involved matter that I cannot discuss in detail here. However, let me briefly sketch its essential aspects. Suppose you have a theory T determined by a class of models M, each of them having a structure of the type

$$m = <D_1,...,D_n,(A_1,...,A_p),R_1,...,R_q>$$

where the D_i are variables for each model's base sets (the theory's "universe of discourse"), the A_j are variables for auxiliary base sets (mostly sets of numbers or similar "purely mathematical" entities), which in some theories may be missing, and the R_k are relations or functions defined over some of the base sets—they represent the theory's specific empirical relations or magnitudes. What is common to all elements of M is that they have the same type of structure as given by m and satisfy the same set of axioms.

Now, suppose that, for whatever reasons, you think that a particular field of experience, call it F, that you (or your colleagues in the "scientific community") are confronted with, can be fruitfully *subsumed* under a structure of type m, satisfying the appropriate axioms. In order to have some chance of success in your endeavor to subsume F under T, the whole process of subsumption must consist of at least three different partial operations:

F1) You "re-conceptualize" F in (some of the) terms of T, which means that you interpret F as a *substructure* d of a particular structure of type m. You may say that you have reconstructed F now as a "model of data" for T^3.

---

[2] I understand the term "model" here in the precise sense of formal semantics, i.e. a model is a structure consisting of some domains of entities and relations defined over them and satisfying certain conditions.

[3] As far as I know, the first author to have introduced this way of speaking was the precursor of the semantic and/or structuralist view of science, Patrick Suppes—see his article Suppes 1962.

F2) By adequately choosing particular values of the variables $D_1,...,R_q$, you *expand* the substructure d into a full structure m. Now you have structured your field of experience F as a kind of structure that *could be* a real model of T, that is, you have it now as a potential element of M. To check whether it is not only a potential but also an actual element of M, you have to proceed to the third operation:

F3), which consists in finding out whether the chosen m really satisfies T's axioms, i.e. whether the chosen values of the variables $D_1,...,R_q$ really cohere so as to constitute a model of T^4.

Now, suppose the operations *F1)—F3)* have already been performed successfully. What are the ontological implications of this? The answer is quite simple: you are now entitled to say that your experience confirms that the "real things behind" your experience are of the kinds settled in the domains $D_1,...,D_n$ of the models of M adequately subsuming the given field of experience F. $D_1,...,D_n$ represent the theory's "ontological commitments". This, in turn, has an almost immediate implication for the formal explication of complexity.

Suppose, indeed, that you suspect that a given kind of thing C is more complex than another kind D. Now, for this statement of a relationship between C and D to make sense, it has to be ensured that these two classes are really comparable. According to the metatheoretical point of view taken here, this means that we have to compare the theories where C and D, respectively, appear as basic domains; and this, in turn, can only be done if these theories subsume the same, or a similar, field of experience. Otherwise the comparison would be spurious. Call these two theories "T_C" and "T_D", respectively. We have then to presuppose that there is an empirical field F which can be successfully subsumed both under models of T_C *and* T_D in the precise sense of the operations *F 1)—F 3)* mentioned above[5]. Consequently, the naïve question about the comparison of complexity between C and D turns out to be a question about the intertheoretical relationship between T_C and T_D; more precisely, it is a question about comparing those models of both theories which subsume the same field F. Let's call such models of two

----

[4] More details about this notion of subsumption of experience under a theory will be found in Moulines 2000.

[5] We could also deal with the slightly more complicated case where the intended fields of experience of both theories are not exactly the same but overlap to some degree, or are in a specifiable relation of similarity, or in some other more or less complicated correlation; but to simplify the discussion here I assume the case where the intended fields of experience are strictly identical.

different theories "empirically comparable". For the rest of this paper, I will always implicitly assume that, for the issue of complexity comparison, we are considering empirically comparable models. Can we make sense of a comparison of degree of complexity with respect to such models? Of course we can! Remember that models are just structures, and there are several, quite plausible, criteria we can apply to claim that one structure is more complex than another. A very simple criterion is that the domains of the first structure are constructed over the basic domains and/or relations of the other. A further criterion is that the first structure requires a more complicated mathematical apparatus to be characterized than does the other. A third criterion is that the first structure satisfies more numerous and/or syntactically more complicated axioms than the second. There might still be other relevant criteria of comparison of structures in this context. Which criteria one uses might also depend on the particular discipline one is considering. My purpose here is not to give a complete and fixed list of such criteria but rather to point out to the fact that, by proceeding in this way, we can develop the most precise and operationally applicable method to decide, by means of structural comparison, when T_C is more complex than T_D, and *a fortiori* when the systems studied by T_C, of which C is a component, are more complex than the systems studied by T_D, of which D itself is a component.

Quite often in the history of science, the following situation has arisen. We have two theories, T and T', such that both subsume a field F more or less equally well as far as successful predictions and tolerable approximations are concerned; but T' covers a field F' of which F is a proper sub-field. At the same time, scientists would like to go on working with T for the restricted domain F for pragmatic reasons (the calculations with T might be simpler, T might be more intuitive, or more elegant, etc.). In such cases scientists will make considerable efforts to prove to themselves and to the rest of the world that T is *reducible* to T'. Now, reduction as an intertheoretical relation is quite an involved matter which I cannot discuss in detail here. However, there are at least two necessary conditions on which there seems to be general agreement. The first might be called the "*concepts-connecting condition*"; the second is the "*laws-connecting condition*". Within our model-theoretic setting, it is convenient to formulate these two conditions in the following way: 1) the basic domains of the models of the reduced theory can be reconstructed as echelon-sets (in the sense of Bourbaki 1968) out of some or all basic domains of the models of the reducing theory. (I remind you that a set A being an echelon-set over the sets $B_1,...,B_n$ just means that A can be reconstructed by successive application of the set-theoretical operations of Cartesian products and power-set building on the sets $B_1,...,B_n$, i.e. A satisfies the condition: $A \in \wp^{(m)} \cdots\cdots \wp^{(n)}(B_1 \times \cdots\cdots \times B_n))$. 2) the actual models of the reduced theory are isomorphic to a specific subset of the set of models of the reducing theory subsuming the same field of experience. I cannot argue here for the rationale for these formulations of the reduction conditions. This has been done at length elsewhere (see e.g. Moulines 1984, Balzer, Moulines, Sneed 1987, Moulines 2000, Moulines 2001). Suffice it to say that, on the one hand, the model-theoretic formulation of these conditions is completely precise and, on the other hand, the first condition, within this model-theoretic frame, can be interpreted as an *ontological* condition: by being reconstructed as echelon-sets over the domains of the reducing theory, the domains of the reduced theory are ontologically derivative—they don't represent independent ontological commitments, though they are still ontological commitments of a sort. To keep clearly in mind what is involved in these two conditions of reduction, let's describe the first one as the condition of *ontological reduction*, while the second might be described as the condition of *nomological reduction*. A necessary condition for a theory T to be reducible to a theory T' therefore is that T be both ontologically as well as nomologically reducible to T'.

Now, consider the following situation, which *prima facie* has some resemblance to reduction as specified here but which on a closer look appears to be quite different. Suppose we have two theories T and T' as before, so that T subsumes an empirical field F which is a proper subfield of the total field that T' subsumes. Suppose, further, that T satisfies the requirement we have called ontological reduction with respect to T' but not the requirement of nomological reduction; quite the contrary, the laws of T are much more accurate and have a higher predictive power than the laws of T' when applied to subfield F. In this situation, though the systems studied in T are, ontologically speaking, derivative with respect to those studied in T' (since they are constructed out of the latter as echelon-sets), they have a kind of nomological autonomy—they can be explained much better by assuming laws that are logically independent of those of the base theory T'. An everyday example might illustrate this situation: undoubtedly, the motion of automobiles can be successfully subsumed under the laws of classical mechanics, in particular when they are moving on a road without obstacles. But when a great number of automobiles are involved in a traffic jam, the resulting system, though ontologically reducible to the single cars, is best subsumed under a theory different from classical mechanics and not reducible to it—rather an application of chaos theory to hydrodynamics. In this case, we might want to say that the traffic-jam system *emerges* from the systems constituted by the individual automobiles. Though the traffic-jam system is ontologically linked to single cars, it has an autonomy of its own, since it satisfies laws different from those of classical mechanics. More exactly, the models of the theory representing traffic-jam systems are neither identical nor isomorphic to a subset of the actual models of classical mechanics.

We hereby obtain a first precise notion of emergence as an intertheoretical relation. A theory T is in the *emergence relationship* to a theory T' whenever the following conditions are satisfied:

E1) The field of experience F subsumed by T is a proper subfield of the field corresponding to T'.

E2) T is ontologically reducible to T'.

E3) T is not nomologically reducible to T'.

In a derivative way, we can then say that the systems treated by T emerge from the systems treated by T'.

We may call this notion of emergence "weak emergence" to distinguish it from a stronger form of this relationship that could also be explicated within our frame. In the stronger form, not only the nomological but also the ontological link between T and T' completely disappears. That is, T is *strongly emergent* with respect to T' when *E1), E3)* and the negation of *E2)* are satisfied. In this case, the systems considered by T are "completely new" with respect to the systems considered by T', though the field of experience the first sort of systems represent is a subfield of the field represented by the second. As far as I can tell from the literature on emergence I know of, some authors have an idea of emergence which rather corresponds to what I have called "weak emergence", whereas others would rather think of strong emergence. At any rate, both notions could and should be clearly distinguished.

What is the relationship of these notions of emergence to the notion of complexity? If the present analysis is accepted, it appears that this relationship is more involved and under-determinate than our intuitive principle, according to which emergence implies a higher degree of complexity, would lead us to expect. In the case of weak emergence, we have a complicated situation. On the one hand, it seems plausible to say that the systems of the theory T, which is in the weak emergence relation with T', are more complex than those of T', since, by the condition of ontological reduction $E2)$, the basic domains of T are constructed as echelon-sets over the basic sets of T', and it is quite natural to say that a set which comes out, in the last analysis, as a successive application of Cartesian products and power-set building to a series of given sets, is more complex than the latter. On the other hand, it could also be the case that the axioms satisfied by the models of T are simpler than those of T'. More complex systems could be shown to satisfy simpler laws. Nothing precludes this possibility a priori. Were this the case, we should say that T is, in a sense, more complex, and in another sense, less complex than T'. Of course, it could be the case, as a matter of fact, that whenever a theory is more complex in the first sense, it also turns out to be more complex in the second. However, such a correlation would only be discovered after a careful analysis of case studies. No general argument can prove it. It is a sort of "meta-empirical hypothesis".

As for strong emergence, any general correlation between it and complexity is still more dubious. Nothing in conditions $E1)$, $E3)$ and non-$E2)$ implies anything whatsoever about greater or lesser complexity. We can only tell that the theory dealing with the strongly emergent systems represents a smaller field of experience than the one of the more general theory. But why should this have any implication for the notion of complexity?

I conclude: the present metatheoretical analysis shows that a correlation between complexity and emergence can only be stated in some particular cases, and that it is, at any rate, conceptually more tenuous than we might have thought at the beginning.

References

Balzer, W., Moulines, C. U. and Sneed, J. D.: 1987, *An Architectonic for Science*. Kluver: Dordrecht.

Moulines, C. U.: 1984, "Ontological Reduction in the Natural Sciences", in W. Balzer, D. A. Pearce and H. J. Schmidt (eds.): 1984. *Reduction in Science.* Kluver: Dordrecht, , pp. 51-70.

— 2000, "Ontologie, réduction et unité des sciences", *Philosophie* 68, pp. 3-15.

— 2001, "Ontology, Reduction, and the Unity of Science", in: Tian Yu Cao (ed.): 2001, *The Proceedings of the Twentieth World Congress of Philosophy* 10, Bowling Green, pp. 19-27.

Suppes, P.: 1962, "Models of Data", in E. Nagel, P. Suppes and A. Tarski (eds.): 1962, *Logic, Methodology and the Unity of Science.* Stanford, pp. 252-261.

Bernulf Kanitscheider

Dept. of Philosophy, University of Giessen, Germany

4. Beyond Reductionism and Holism. The Approach of Synergetics

Since time immemorial, advocates of reductionism and defenders of holism have disputed over the question whether systemic qualities of complex structures can be understood on the basis of their elements and the interactions among them. Reductionists take holistic thinkers to be dark and nebulous; holists consider adherents of reductionism to close their minds to the emergent qualities of composite systems.

The tenor of today's thinking, however, is set by the partisans of analytic methods, who with their dissecting methodologies have succeeded in pushing back, at least apparently, those who defend the idea that there exist entities resistant to analysis.

"The love of complexity without reductionism makes art; the love of complexity with reductionism makes science"[1]. With these words, Edward Wilson expresses the conviction of many "working scientists". Steven Weinberg's opinion is similar: Reductionism is a useful filter that prevents scientists in all domains from wasting their time with ideas that are not worth the trouble[2]. And P. Bak applies economic terms: "The reductionist approach has always been the royal road to the Nobel Prize"[3].

It has happened more than once in the history of science that a philosophical controversy has been brought closer to a solution by a special scientific theory. In many cases the philosophers of the opposing parties were both wrong-headed in their ideological trench-warfare in not being able to see a third conceptual alternative. With regard to the discussion

---

[1] Wilson 1998, p. 54.

[2] Weinberg 1993, p. 71.

[3] Bak 1997, p. 114.

concerning what have been termed emergent systemic properties—which were denied neither by the reductionist party nor by the holistic side, both being eager to put them to work on their own behalf—a new scientific discipline has drawn attention to itself: synergetics.

This new science, founded by Hermann Haken, goes far beyond the sterile antagonism between reductionism and holism, proposing a new view on interactive systems. This science cannot be reduced to any of the existing special sciences, operating as it does at the meta-level of the autonomous organization of complex systems with arbitrary material properties, and not being linked to any particular ontology. Naturally, it must realize, step by step, its claim to being able to formulate the laws according to which complex arrangements in nature and society are composed. At present, we are not sure whether synergetics is the definitive universal theory of the spontaneous organization of structural and functional order, just as, in 1687, it was not known whether Newton's theory of gravitation applied to the whole universe. After all, the reach of a conceptual scheme can only be determined a posteriori.

If synergetics claim to be able to construct a comprehensive scheme of morphogenetics is correct, we can, indeed, speak of a new bridge connecting the natural, social and psychological sciences, the connecting principle being the unitary description of co-operative emergent phenomena.

Synergetics was originally a theory of a special science. It was anchored in quantum optics, in which it depicted the laser, with its well-formed non-linear differential equation and the conspicuous effect of monochromatic laser light. However, about 20 years ago the protagonists of synergetics transcended the level of physics, suggesting that the synergetic co-operation model could be amplified, or even universalized[4]. Many applications were demonstrated in which e.g. instability, phase transitions, the bifurcation of lines of development, and the creative potential of chaos played a key role in the formation of structures immanent to a system. To date, the analysis of the complete set of proposed applications has not yet been achieved in a satisfactory manner; and the question of the reach of the explanatory power of the theory, above all in the social sciences, is probably still open[5]. It already seems clear however that the metatheoretical controversy between reductionism and holism had simply focused on complementary aspects of a process that is ultimately unitary, which means that the whole approach was

---

[4] See Haken 1990.

[5] Weidlich and Reiner 1994, p. 177.

actually incomplete. Both positions had concentrated their energies exclusively on the opposite elements within the process of emergence: *reductionism* focused on the combination of microscopic parts, while *holism* considered new qualities at the macroscopic level. Synergetics, however, links both levels. Its success is due to three concepts: parameter of order, principle of enslavement and circular causality. The macroscopic variables which characterize the emergent phenomena and those describing the individual parts of a system are causally interrelated. The elements of the system generate their own organizers, and these variables react on the parts determining their behavior by forcing them into particular modes[6]. The famous picture by M. C. Escher, where two hands are drawing each other, could be taken as a nice illustration of the two levels interacting according circular causality.

It is easy to see that the advantage of synergetics consists in an elegant simplification by which a complex non-linear system can be described in terms of ordering parameters. Moreover, the structural properties of phenomena can be understood following the causal process of their emergence: they do not appear in a sudden, miraculous way, like entities falling from heaven. They can rather be conceived according to the dynamics and co-operative behavior of a non-linear system. This approach is by no means reductionist, because there is no question of reducing the co-operating phenomena to an elementary unitary force, a strategy which has been advanced by theoreticians of unification, for instance in quantum gravity and in the superstring approach. If we can speak of a striving for the unity of science here, it is not in the sense of a unitary theory of all interactions, but in that of a structural transdisciplinary common ground of all structure-generating processes. As synergetics claims universal validity, there is, of course, a considerable risk of failure. But, on the other hand, if synergetics did not risk failure through the discovery of some emergentist phenomenon, e. g. in social or cultural reality, it could not be characterized as a scientific cognitive enterprise.

The conditions of application can be specified rather clearly: a thermodynamically open non-linear system must be involved, which evinces an action field for arbitrary fluctuations in such a way that under certain boundary conditions collective patterns of motion enter into competition with each other which, after a phase of instability, leads to a new long-lasting form of motion. It is a matter of empirical investigation to ascertain

---

[6] See Haken and Knyazeva 2000.

whether these assumptions which are necessary to commence the process of self-organization are present in a particular case. In a way, synergetics is rather a frame theory, that is to say, an abstract theory-scheme which, similarly to classical mechanics, will become a definitive theory only after having specified a force function.

Some critics of synergetics have remarked that numerous transferences of scientific patterns of thinking to social and literary domains are nothing but more or less vague analogies and metaphorical ways of speaking. But we should not forget that analogies, in the context of theory construction, often have a fruitful, knowledge-guiding heuristic function. But, at the same time, it has to be admitted that phenomena of social reality quite frequently do not have sufficient conceptual sharpness to allow the application of a metrical language. For example, how is it possible to clarify the self-organization of thought in the human brain to the point where the formal apparatus of synergetics can be employed? On the other hand, we don't see why, at a certain level of evolutionary complexity, the explanatory model should suddenly fail simply because the nature of the constituents has changed. At least in the case of language the synergetic order-scheme seems to be applicable in a rather plausible way, when we consider the fact that individual speakers are enslaved by the speech community, while they themselves constitute that very community. And the determining reaction of the native language group upon the speech habits of individuals perfectly fits the idea of the enslaving principle, which does not have an ethical dimension in any sense.

One of the decisive conditions of the applicability of synergetics is the non-linearity of the system being investigated. The system must allow positive feedback in fluctuations as a starting point for macroscopic changes. This formal property is linked, as well, to a threshold of sensation. Only after having exceeded this limit—e.g. if in the intellectual geography of the scientific community a special idea has been sufficiently widely accepted—can the system instigate the growth of structure. In the opposite case, where the system ignores the fluctuation, it becomes drowned in the river of transitoriness, *"en el flujo de lo corruptible"*, according to Ortega. Non-linear systems are, moreover, characterized by a particular variety of discrete evolutionary branches which represent a factor of discontinuity when emergent phenomena appear. Non-linear systems, when taking a particular path at a bifurcation point, show an arbitrary behavior which is not due to quantum mechanics, and this characteristic leads to a philosophically important consequence: these systems are unpredictable, in that their behavior at the very point of bifurcation cannot in any way be extrapolated into the future. If we apply this trait of the theory to cosmic development, we are led to a conclusion which seems to have some theological importance: e.g. an extramundane demon of the Laplacian type is unable to foresee what kind of systems will develop in a universe whose evolution is steered by instabilities, nor what kind of qualities these systems will manifest. According to synergetics the self-structuring of a non-linear dynamic system occurs in a restricted indeterministic manner, which means that not all conceivable paths into the future are possible. The evolution of the universe takes place within a discrete spectrum of alternatives. The protagonists of the synergetic approach formulate their long-term goal following the program of unitary theories. But in this case it is not important to find the fundamental field which is able to provide the classes of particles supplied by experiment. The ideal is rather to establish a class of non-linear equations which can describe the pattern of possible paths of evolution as well as the stable conditions of self-organizing systems. In such a way, synergetics aims at the revival of an old Ionian dream: to understand the variety of the world, its spatial and temporal structures, on the basis of a unitary building plan. In contrast to Presocratic ideas which searched mainly for a static ἀρχή, it benefits from including the evolutionary element. Therefore, it not only aims at self-organization, i.e. the structure or the topology of macroscopic organization, but also makes statements on the velocity with which coexisting patterns develop, and on their system-dependent proper time. In any case, synergetics casts a new light on the old aporetical philosophical controversy between reductionism and holism, and gives substantial power to the vague idea of interdisciplinarity. Once again, it shows that philosophy and science are not divided by a sharp border and, moreover, that new scientific approaches are able to resolve old philosophical controversies of philosophy, or dissolve them as rash alternatives.

References

Bak, P.: 1997, *How Nature Works?*. Oxford.
Haken, H.: 1990, "Synergetik und die Einheit der Wissenschaft", in W. Saltzer (ed.): 1990, *Zur Einheit der Naturwissenschaft*. WBG Darmstadt, pp. 69-78.
Haken, H. and Knyazeva, H.: 2000, "Synergetik: Zwischen Reduktionismus und Holismus", *Phil. Nat.* 37, 1, pp. 21-44.

Weidlich, W. and Reiner R.: 1994, "Der Beitrag der Synergetik zum Naturverständnis", in G. Bien and T. Gil: *Natur im Umbruch*. Stgt.: Frommann.
Weinberg, S.: 1993, *Der Traum von der Einheit des Universums*. München.
Wilson, E. O.: 1998, *Consilience. The Unity of Knowledge*. New York.

5.   Kolmogorov Complexity

Jesús Mosterín

Dept. of Philosophy, CSIC, Madrid, Spain

1. Vagueness of the everyday notion of complexity

The word 'complexity' is frequently used in a vague and undefined way. Judgments of comparative complexity are being made all the time, but seldom on the basis of any explicit criterion.

Sometimes complexity—what P. Godfrey-Smith calls "complexity as heterogeneity" and Rescher calls "compositional complexity"—is supposed to be related to the number of different components and their interrelations. When we say that our brain is very complex, we can be referring to the high number (to the order of 1011) of neurons it is composed of and to the many connections among them. But consider a typical galaxy like the Milky Way. It has a similar number (1011) of component stars. And all stars are gravitationally connected to each other. Of course, both systems are far from being fully understood. Which is more complex, a brain or a galaxy? Some people would answer that a brain is more complex. But why? According to what criterion?

We often hear that a certain organism is more or less complex than some other. Is a man more complex than a mouse? Is a dog more complex than an onion? Why? To measure the absolute complexity (whatever that may be) of a real thing like an animal seems to be a task of intractable difficulty. In order to make the task more amenable to precise treatment, some theoreticians have advised not looking at the thing itself, but at the description of the thing in some standardized language or means of representation.

The complexity of an organism, for example, would be the length of its description. The most accurate description of an organism seems to be its own genome. So the first idea was to define the complexity as the quantity (the total number of bases) of its DNA per cell. But we were not pleased to learn than amphibians have more DNA per cell than mammals (including humans). Humble onions have five times more DNA per cell than we have, and tulips have ten times more! The second idea was to discount the introns and the repetitive sequences of DNA and, in general, the non-coding so-called junk DNA, and to concentrate instead on counting the genes. But we have learnt in 2001 that humans do not have 100,000 genes, as previously assumed, but just 30,000, the same number as mice. It seems that all mammals have a similar number of genes in their genome. Perhaps we have more complicated regulatory rules. Perhaps, but we do not really know. We will see.

How can we make the notion of complexity precise? Kolmogorov's idea was to simplify the problem: instead of asking for the complexity of the thing, ask for the complexity of its description. And reduce the qualitative notion of complexity to the quantitative notion of size. The complexity of something is the size of its shortest description. The shortest description is the shortest program that generates the whole description in a certain computer, a universal Turing machine. And the size of this description is just the length of the program. More on this later.

2. The compression of coded information

Coded information is able to be more or less efficiently coded. An inefficient, or redundant or too long code can be made shorter, more efficient or compact; it can be compressed. Organic nature makes use of this compressibility of information. The measures of the structural complexity of an organism (even of our brain) give much higher values than the structural complexity of the DNA of the organism. The information for building the organism has been greatly compressed in the DNA codification. On the other hand, we know that DNA encoding is far from being optimally efficient. Not only is the code itself redundant, but the particular genomes contain further redundancy in the form of multiple repetitions of the same DNA segments.

The more complex a text is, the less compressible it is. A text x is less complex than another text y if and only if x can be generated by a shorter algorithm or program than y. The text of a poem or a song can include several repetitions of the same stanza or refrain, in which case it is possible to spare letters in the specification of that text. It suffices to write the repeated stanza only once and to indicate each of its repetitions by a short mention. If any long name is repeated in the text, it is also possible to shorten it after its first occurrence, with a new saving of characters. So it is often possible to completely specify a text of n characters by means of only m characters, where m < n. The shortened text contains the same information as the complete text, but the shortened text codifies it more compactly, in fewer letters. This shortening process achieves a compression of the information. The more regular, repetitious or symmetrical the text is, the more suitable it is for compression. The more irregular it is, the less amenable to compression it will be. This resistance to compression can be taken as an indicator of complexity.

The compression of information has been a pressing concern in the recent development of integrated audio-video computer systems, digital cameras and video recorders, and interactive web services. These systems require everything (images, sound, data) to be stored in digital form. The problem was that sounds, pictures and, especially, images in motion—motion pictures—require too many bits of memory for the capacity of the previous storage systems, such as CD-ROMs, at least if the usual encoding is kept. In the monitor of a computer or the screen of a TV set a still image is represented as a frame of small color points called pixels. The number of pixels depends on the resolution of the screen. A low resolution screen can have—let us say—one million pixels. If the system chooses from a palette of—say—256 colors, we need 8 bits (= 1 byte) for the choice of the color of each pixel, as $\log2(256) = 8$. If we add another byte for the intensity value, we get 16 million bits (= 2 million bytes) as the space needed to store the information of a still frame.

In order for the human eye to perceive the impression of continuous movement, we need to show something like 25 still frames per second. That means the memory space needed to store one second of video is 50 MB. The standard digital information device used to be the CD-ROM. But a CD-

ROM has a capacity of about 750 MB, and that is enough for only 15 seconds of low-resolution video. So we see how daunting the difficulties were for the developers of integrated digital video systems. The solution has been found in the compression of information. The more recent technology of the DVD allows for the recording of seven times more bytes than the CD in each of the two layers on each side of the disk, so that a total of 17 GB of data can be stored in one DVD. Still, without the help of information compression, this huge increase in storage capacity would only allow six minutes of video. With compression, it is enough for a whole movie of two hours duration.

The still frames of a movie often contain homogeneous zones, for example in the background of the picture. The pixels of those zones can be economically specified by default, assuming that all pixels not specifically described are—say—of a dark gray color. High levels of compression can also be achieved in the encoding of the movement scenes, in which each frame is almost identical with the previous one, with only very slight modifications. Perhaps an arm is slowly moving, the rest of the landscape remaining the same. Then it is enough to code for those changes, giving as instruction for the rest of the frame the repetition of the previous information. With tricks like these it is possible to save huge numbers of bits and to compress the information contained in a movie onto a single DVD. Special algorithms are needed for the compression and the decompression of the images. And in order for these algorithms to run quickly enough for the images to appear smoothly on the screen and in concurrence with the other processes of the system, special co-processors have to be installed in the hardware.

Any text, any amount of data, any image or video, any melody or sound, can be encoded as a binary sequence, i.e., as a sequence of zeros and ones. This is the way compact disks store music and computers store data, texts and pictures, and DVDs store movies. Any coded information can be transcoded into a binary sequence. And the simplicity or complexity of the original message will reappear as the simplicity or complexity of the corresponding binary sequence. So the study of complexity can be restricted, without loss of generality, to the study of the complexity of binary sequences. That is precisely the endeavor of the algorithmic theory of complexity. This theory makes essential use of the notion of a Turing machine.

3. Turing machine

The concept of computability has been analyzed, explicated and made precise in a unique and satisfactory way. As a matter of fact, the elucidation of the notion of computability (or of a computable function) can be considered as the greatest triumph of mathematical logic in the 20th century. We have tried to make precise the notions of consequence, of logical validity and of a set, but we have come out with a variety of non-equivalent definitions. So we have different and irreducible notions of consequence, of validity and of a set. But all the proposed definitions of the notion of computability (by Post, Church, Gödel, Turing, Markov, Kleene and many others) have been proved to be equivalent. Church's thesis has been vindicated. Kolmogorov's idea was to use the extremely precise and successful ideas of computability theory, especially the idea of a universal Turing machine, to give a quantitative definition of complexity.

A Turing machine is an idealized computer, whose input takes the form of a binary sequence on a potentially infinite tape. The workings of a Turing machine are totally specified by a table or matrix, with which the machine itself can be identified.

An m-state Turing machine over an alphabet $\{|\}$ is a 2m×4 matrix (a table of 2m rows and 4 columns) of the form

State	Scanned symbol	Action	New state
1	*	a11	c11
1	|	a12	c12
2	*	a21	c21
2	|	a22	c22
.	.	.	.
.	.	.	.
.	.	.	.
.	.	.	.
m	*	am1	cm1
m	|	am2	cm2

where for each i,j ($1 \leq i \leq m$; j = 1 or j = 2):

the actions $a_{ij} \in \{*, |, R, L, S\}$

the new states $c_{ij} \in \{1, 2, ... m\}$.

This table can also be coded as a binary sequence. Alan Turing proved in 1937 that there are universal Turing machines. A universal Turing machine U is a Turing machine that simulates the behavior of any other Turing machine. Let x be an input (a binary sequence), let T be a Turing machine (coded also as a binary sequence), and let T(x) be the result or output produced by the machine T after processing input x (if such processing ever halts). A universal Turing machine U is a Turing machine, such that for every binary sequence T that codes a Turing machine and every binary sequence x which is a possible input of that machine, U processes such input exactly as the machine T would do it, i.e., for every T and every x (both binary sequences):

$$U(T,x) = T(x)$$

A universal Turing machine, conveniently programmed, is able to compute any computable function and, in particular, can generate any computable binary sequence.

4. Algorithmic Complexity Theory

Let us consider the binary sequences A and B whose first digits are:
A: 001001001001001001001001001001001001001001001001001001001 ...
B: 00101110100100001101010001011111101101010010111010000011 ...

Suppose both sequences are 3 million digits long. Sequence A can be described or generated by means of the simple algorithm: "write 001 one million times". Sequence B does not seem to be describable in a much shorter way than by just specifying the actual sequence in its entirety. The first sequence is highly regular. It is simple. The second one appears very irregular. It is complex.

One exact measure of the complexity of a binary sequence is the length of the minimum program that generates that sequence. Of course that measure would be useless, if it were relative to a particular computer or to a particular programming language. Fortunately it is possible to arrive at an absolute value (up to an additive constant), independent of any variation in the hardware or the programming language. Any universal Turing machine will do the job.

Let us now fix a certain universal Turing machine U. A program for generating the binary sequence x is a binary sequence p, such that, when U receives p as input, it produces x as output, i.e., such that $U(p) = x$. Each of these programs has a certain length (a certain number of digits—zeros or ones): length(p). The minimum length of such programs is a precise measure of the complexity of the binary sequence x. If no program generates x, we say that the complexity of x is infinite, ∞.

This measure is unique up to an additive constant. If, instead of having chosen the universal Turing machine U, we had chosen a different universal Turing machine, let us say U0, then the different resulting measures of complexity would have coincided asymptotically, i.e., the difference of their values would have always been less than a fixed number c (which depends only on U0), so that for sufficiently long sequences, both measures would have practically coincided. (A more precise statement of this fact is called the invariance theorem of complexity theory).

$K(x)$, the complexity of a binary sequence x, is the length of the minimal program p that generates x, if any programs generating x exist, and is ∞, otherwise.

$$\exists p \, U(p) = x \Rightarrow K(x) = \text{the minimum n such that } \exists p[n = \text{length}(p) \wedge U(p) = x]$$

$$\neg \exists p \, U(p) = x \Rightarrow K(x) = \infty$$

The function K is not computable, but has computable approximations.

The oldest precedent of algorithmic complexity theory can be traced to von Mises' attempts to define the notion of a random binary sequence during the period between the wars. The concepts and ideas typical of the theory appeared for the first time at the beginning of the 60's in the work of Ray Solomonoff. Finally in 1965 Andrei Kolmogorov published "Three approaches to the quantitative definition of information", where he precisely defined the notion of complexity—now called in his honor Kolmogorov complexity or K—as a measure of the randomness of binary sequences, and proved the invariance theorem. Other mathematicians, like G. Chaitin, P.

Martin-Löf, L. Levin, P. Gacs and D. Loveland, have also contributed to the development of the theory[1].

Shannon's concept of informational entropy is based on the existence of a set of possibilities or alternatives, provided with an a priori probability distribution, and measures our ignorance of which of these possibilities has materialized. Solomonoff, Kolmogorov and Chaitin, on the contrary, are interested in the informational content of an individual object, without any reference to a set of alternatives. They define the complexity of that object in terms of the difficulty of describing or generating it.

If an object is very regular, it is easy to describe, and so it is simple. If, on the contrary, it is very irregular, then it is difficult to describe, and so it is complex. If it is so complex that the information it contains cannot be compressed, we say it is random. So randomness is characterized as the maximum of complexity, and as the opposite of regularity and simplicity.

5. Mathematical description as an encoding process

For information to be defined, we need a well-defined framework, with clear-cut alternatives. Mathematical description can provide such a framework. Placed into such a framework, a raw chunk of reality becomes a system and yields information. It is this rigid framework that creates the conditions for coded information to arise in the first place.

The raw information present in reality has been made available as coded or usable information through the simplifying, idealizing and clarifying process of mathematical description and model building.

In order to obtain theoretical knowledge of the real world, we force it into the mold of some mathematical structure. The real shape of the Earth is ineffable. But we think of it as a sphere, we model it as a sphere, and so we are able to ask for its radius, and to compute its surface area and volume. In successive approximations, we can project more complicated geometrical forms (such as a spheroid of revolution) onto it and get new and more accurate information.

The mathematical world is fictitious, but objective, well defined, with its own truth, with its clear sets of alternatives, on which to project the real, but

---

[1] For a good summary of the theory, see Li and Vitany 1997.

fuzzy world of experience. The raw information present in observation has to be filtered, smoothed out and clad in mathematical or theoretical form in order to become coded information, observational report.

6. Axiomatization as compression

A law or formula compresses the information contained in many observations and historical data.

Solomonoff, a disciple of Carnap, shared his teacher's interest in induction, but looked at the subject with a fresh eye. He pointed out that there is a lawlike relationship among a set of observations if and only if the series of their descriptions is not random, i.e., if their regularity makes them compressible. He introduced a general theory of inductive reasoning, based on the idea that a scientific law represents a particularly efficient way of compressing the information present in many observational reports, which, once coded as binary sequences, are conceived as the initial segments of an infinite binary sequence, generated by the law. This account is relatively unproblematic for low-level generalizations and phenomenological laws. And, given suitable precautions, it can be extended to laws and theories in general.

The axiomatic method is a method for the compression of information. The axiomatization of a theory is the most efficient way to encode the information contained in its theorems. In some cases each theorem can be conceived of as a law, summarizing many real or possible observations. When we compress the information contained in the observational reports, we get the laws. And when we compress the laws, we get the axioms. The axioms can then be thought of as laws of laws, as more efficient encoders. The shortest independent axiom system is then random (if it were not, it would not be the shortest one).

We can think of a theory as a way to compactly summarize all the various sentences it is able to prove, as a program that strongly compresses the information contained in the infinite set of its theorems. The diverse consequences of a theory are encoded by the theory's axioms (and the underlying logic). As the axioms are usually finite in number and short in length, and the consequences are infinite in number and of any length, the compression achieved through successful axiomatization can be stupendous indeed.

7. Complexity and the limits of compression

Each binary sequence represents a natural number in the base-two numeration system. We can define the complexity of a natural number as the K-complexity of its base-two representation. Most binary sequences and natural numbers are very complex, and in this sense they can be said to be random. Even if it is easy to prove that a sequence is not random and thus that its information is compressible (it is enough to present the corresponding compact algorithm which generates it), it is difficult or impossible to prove that a particular sequence is random or incompressible. So, we can prove that most numbers are random, but are unable to prove that a particular random number is random.

Most numbers are random and, as they become larger, their complexity grows over all finite bounds. But in a theory of a given degree of complexity it is impossible to prove that a number (or its corresponding binary representation) has a complexity much greater than the complexity of the theory itself. In fact, it is possible to associate with any consistent formal theory which includes elementary arithmetic a constant (a natural number) c-—which depends on the theory—such that no proposition of the form "$K(x) > c$" (where x is a finite binary sequence and K is the complexity) can be proved in the theory. In this sense the theory of computational complexity allows us to obtain incompleteness results similar to those of Gödel[2].

8. Alternatives to K-complexity

Kolgomorov's measure of complexity, K-complexity, has several impressive advantages: it is objective, mathematically well defined, and fruitful in theoretical applications. It also presents important failures: K is not computable in general. And it departs from the intuitions of some people, who object to the identification of complexity with randomness. They would

---

[2] See Chaitin 1992 and van Lambalgen 1989.

like complexity to be an interesting middle point between the two uninteresting extremes of extreme K-simplicity and extreme K-complexity (randomness). But "interesting" is a subjective category, difficult or impossible to define in an objective way.

Shannon's notion of informational entropy is an alternative to K-complexity, but both theories contain similar theorems. Kolmogorov's has the advantage of being applicable to concrete single objects, and not just to ensembles, like Shannon's.

Time is irrelevant in Kolmogorov's complexity theory. What is important is only the program size, not for how long the program runs. Once we introduce time bounds, the theory gets much more complicated and murky. The nice theorems (which parallel Shannon's theory) no longer hold.

Charles Bennet has introduced the notion of logical depth. K-complexity is a program-size complexity, it does not take into account the difficulty, or the time, or the number of steps of the actual computation. We can have two different programs of similar length, but one of them runs for only a few steps, whereas the other performs long and intricate computations before coming to a halt. The logical depth of the second is much greater than that of the first. The logical depth of an object (a binary sequence) is the time (the number of steps) required by the universal Turing machine to compute or generate it from its maximally compressed description.

Murray Gell-Mann has introduced the notion of effective complexity, by which he means the length of a very concise description of the regularities of a system or binary string. So effective complexity measures only the regularities of the system, ignoring everything that is random in it. It is the number and variety of regularities that make up effective complexity. Nevertheless the difference between the regular and the random, between the signal and the noise, is context-dependent and observer-dependent. The distinction depends on a judge who decides what is important or not, what is signal or noise, what counts as regularity. If we eliminate the judge, we fall back onto Kolgomorov's complexity. If we allow for the judge, we can have a more intuitive notion of complexity, but at the price of losing the objective and absolute character (up to an additive constant) of Kolmogorov's measure, and having to accept observer-dependency and relativity.

Anyway, it is clear that we use the word 'complexity' in a wide variety of ways and associate it with diverging intuitions. It is not to be expected that any formal specification of the notion will ever capture all these nuances of ordinary usage and intuition. Kolmogorov's measure is no exception, but at least it has the double virtue of existing and of being mathematically well defined. So, when we talk about K-complexity, at least we know what we are talking about.

References

Chaitin, G.: 1992, *Information, Randomness & Incompleteness*. New York: IBM.
Cover, T. and Joy, T.: 1991, *Elements of Information Theory*. John Wiley & Sons, Inc.
Gammerman, A. and Vovk, V.: 1999. "Kolmogorov complexity: Sources, theory and applications", *The Computer Journal* 42, pp. 252-255.
Gell-Mann, M.: 1994, *The Quark and the Jaguar: Adventures in the Simple and the Complex*. Little, Brown and Company.
Godfrey-Smith, P.: 1996, *Complexity and the Function of Mind in Nature*. Cambridge University Press.
Kolmogorov, A.: 1965, "Three Approaches to the Quantitative Definition of Information", *Problems in Information Transmission* 1, pp. 1-7.
— 1968, "Logical Basis for Information Theory and Probability Theory", *IEEE Transactions on Information Theory* IT-14, pp. 662-664.
Li, M. and Vitanyi, P. (eds.): 1997, *An Introduction to Kolmogorov Complexity and its Applications* (2nd ed.). New York: Springer Verlag.
Mosterín J.: 1992, "Theories and the Flow of Information", in Echeverria, Ibarra and Mormann (eds.): 1992, *The Space of Mathematics*. Berlin: Walter de Gruyter, pp. 366-378.
Rescher, N.: 1998, *Complexity: A Philosophical Overview*. Transaction Publishers.
Shannon, C. and Weaver W.: 1949, *The Mathematical Theory of Communication*. University of Illinois Press.
Solomonoff, R.: 1964, "A formal theory of inductive inference" Part I and II, *Information and Control* 7, pp. 1-22 and 224-254.
Turing, A.: 1937, "On computable numbers, with an application to the Entscheidungsproblem", *Proc. London Math. Society* 42, 2nd series, pp. 116-154.
Uspensky, V.: 1992, "Kolmogorov and Mathematical Logic", *Journal of Symbolic Logic* 57, pp. 385-410.
van Lambalgen, M.: 1989, "Algorithmic Information Theory", *Journal of Symbolic Logic* 54, pp. 1389-1400.

6. Modèles de structures émergentes dans les systèmes complexes

Jean Petitot

EHESS & CREA, École Polytechnique, Paris, France

Introduction

Le paradigme des phénomènes d'émergence de propriétés dans les systèmes complexes reste celui de la physique statistique. De l'interaction complexe (coopérative ou compétitive comme dans les verres de spins) d'un grand nombre d'unités "micro" élémentaires émergent des propriétés "macro". Un *changement de niveau* d'organisation résulte d'une telle émergence.

Peut-être que l'un des apports scientifiques les plus novateurs et les plus fondamentaux de ces 30 dernières années aura été de fournir de nouveaux exemples, encore beaucoup plus profonds, de ce phénomène. Il ne s'agit plus seulement de l'émergence de *grandeurs* macroscopiques comme la température, la pression ou l'aimantation, mais de *patterns*, autrement dit de *structures morphologiquement organisées*. Il existe une relation fascinante entre, d'un côté, ces structures émergentes macro qui possèdent une organisation en constituants permettant de les décrire de façon relativement économique et, d'un autre côté, l'étonnante complexité de leur implémentation qui se trouve distribuée sur un nombre énorme d'unités élémentaires.

Je vais me borner à donner 2 exemples très différents, l'un physique, l'autre neurocognitif, concernant l'émergence de *formes* organisées dans des substrats. Mais insistons sur le fait que les systèmes complexes possèdent des propriétés macroscopiques globales émergentes provenant d'interactions collectives coopératives-compétitives. Ils sont *singuliers*, en grande partie contingents (non concrètement déterministes, ils présentent une sensitivité à des variations infinitésimales de leurs paramètres de contrôle, sensitivité produisant des effets de divergence). Ils sont historiques et résultent de processus d'évolution et d'adaptation. Ce sont des systèmes hors équilibre possédant une régulation interne leur permettant de demeurer à l'intérieur de leur domaine de viabilité. Is n'ont plus rien à voir avec le déterminisme mécaniste laplacien qui sert de repoussoir aux critiques anti-positivistes.

1. Vers une physique des formes

C'est René Thom qui a défini le premier de façon à la fois mathématique et générale ce que sont une morphologie et un processus morphogénétique. L'idée fondamentale est de considérer qu'en chaque point w de l'espace W du substrat de la forme il existe une dynamique *locale*, dite dynamique interne, X_w qui définit la physique ou la chimie ou le métabolisme bcal du substrat. Ce régime local, cet état interne du substrat, se manifeste phénoménologiquement par des qualités sensibles (couleur, texture, etc.). Les rapports de voisinage spatial entre les différents points w induisent alors des *couplages* entre les dynamiques internes locales. Celles-ci interagissent et des *instabilités* peuvent donc se produire. Cela entraîne des bifurcations des régimes locaux, des brisures des symétries du substrat, brisures qui entraînent à leur tour des discontinuités qualitatives dans l'apparence du substrat. Et ce sont ces ruptures d'homogénéité qui engendrent enfin les formes. L'idée principale est donc de considérer l'espace et le temps non plus comme un simple contenant pour des objets mais *comme un espace de contrôle* permettant de faire interagir des dynamiques internes locales voisines.

Ce point de vue fournit un cadre théorique unitaire à tout un ensemble de travaux. Je cite deux exemples, celui des modèles de Turing et celui des champs continus d'oscillateurs de Pierre Coullet.

Les équations de réaction-diffusion introduites par Turing en théorie de la morphogenèse permettent de comprendre l'émergence de motifs morphologiques macroscopiques dans les réactions chimiques. Elles couplent des équations cinétiques de réaction décrivant des interactions moléculaires locales et des équations de diffusion décrivant des phénomènes de transport. La diffusion produit de l'uniformisation, elle homogénéise. C'est pas excellence un processus destructeur de morphologies. Mais si le milieu est le siège de réactions chimiques avec catalyse et autocatalyse (les équations différentielles de la cinétique chimique exprimant l'évolution temporelle des concentrations des espèces chimiques sont alors non linéaires) et s'il est loin de l'équilibre thermodynamique (système ouvert) alors il peut y avoir des morphologies spatio-temporelles complexes qui émergent de façon stationnaire et qui sont engendrées par des processus d'auto-organisation. Le caractère explosif de l'autocatalyse se trouve inhibé par d'autres réactifs et, suivant les vitesses de diffusion relatives des produits de la réaction, les morphologies peuvent être très différentes.

Par exemple si A est un activateur auto-catalytique et si H est un inhibiteur dont la synthèse est catalysée par A, alors à partir d'une situation initiale homogène on peut obtenir des motifs périodiques. Une petite fluctuation de A produit par autocatalyse un pic local de A. Mais cela amplifie aussi la concentration de H localement. Mais si H diffuse plus vite que A, la formation de A ne sera inhibée par H que latéralement et non pas au centre du pic. D'où un pic de A bordé par un défaut de A.

Un exemple de système d'équations non linéaires modélisant un tel système sont par exemple :

$$\begin{cases} \dfrac{\partial a}{\partial t} = \rho \dfrac{a^2}{h} - \mu_a a + D_a \dfrac{\partial^2 a}{\partial x^2} + \sigma_a \\[2ex] \dfrac{\partial h}{\partial t} = \rho \dfrac{a^2}{h} - \mu_h a + D_h \dfrac{\partial^2 h}{\partial x^2} + \sigma_h \end{cases}$$

où $a(x,t)$ et $h(x,t)$ sont les concentrations respectives de l'activateur A et de l'inhibiteur H, où les termes non linéaires en a^2 expriment l'autocatalyse de A et la catalyse de H par A, où le terme en $1/h$ exprime l'inhibition de la production de A par H, où les termes linéaires $-\mu_a a$ et $-\mu_h h$ sont des termes de dégradation (les constantes μ sont des durées de vie de molécules et $\mu_a < \mu_h$: H se dégrade plus vite que A), où les termes $D_a \dfrac{\partial^2 a}{\partial x^2}$ et $D_h \dfrac{\partial^2 h}{\partial x^2}$ sont des termes de diffusion avec $D_a \ll D_h$ (H diffuse plus vite que A), et où enfin les termes constants > 0 σ_a et σ_b garantissent que les espèces chimiques A et H restent toujours présentes.

On peut obtenir ainsi des morphologies complexes, par exemple des structures en bandes (structures localement simples mais globalement complexes avec des défauts, des points d'arrêt, des dislocations, etc. comme dans les cristaux liquides). (Cf. fig. 1).

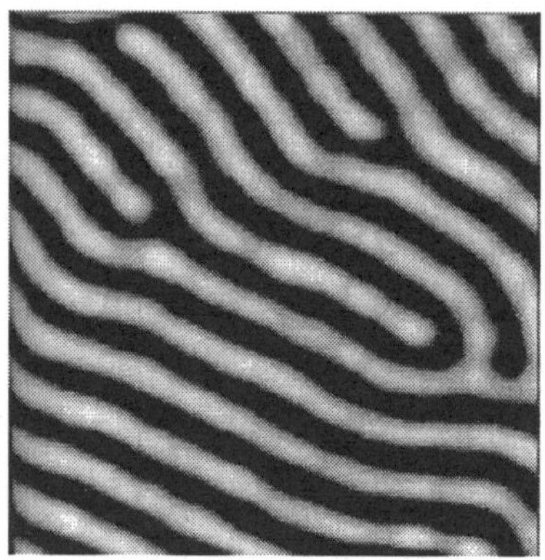

Figure 1: De Kepper.

En analysant les instabilités de champs continus d'oscillateurs, Pierre Coullet a montré comment on pouvait engendrer un nombre considérable de formes de types différents. On considère par exemple des oscillateurs faiblement couplés par leurs relations topologiques de voisinage et soumis à un forcing avec une fréquence voisine du double de leur fréquence propre. La variable locale observée peut être l'amplitude ou la phase de l'oscillateur. L'amplitude de la modulation et l'écart à la résonance sont des paramètres.

En passant à la limite d'un continuum d'oscillateurs dont le paramètre d'ordre (la phase moyenne) Z dépend de la position spatiale, on obtient des équations du type:

$$\frac{\partial Z}{\partial t} = \lambda Z - \mu|Z|^2 Z + \gamma_n \overline{Z}^{\,n-1} + v\Delta Z$$

où λ, μ et v sont des paramètres complexes et γ_n un paramètre réel.

Ces oscillateurs peuvent se synchroniser et se désynchroniser localement. En introduisant de la diffusion, on obtient une très riche variété de patterns spatiaux: turbulence développée, défauts, ondes spirales, cellules hexagonales, réseaux de bandes, etc. (Cf. fig. 2).

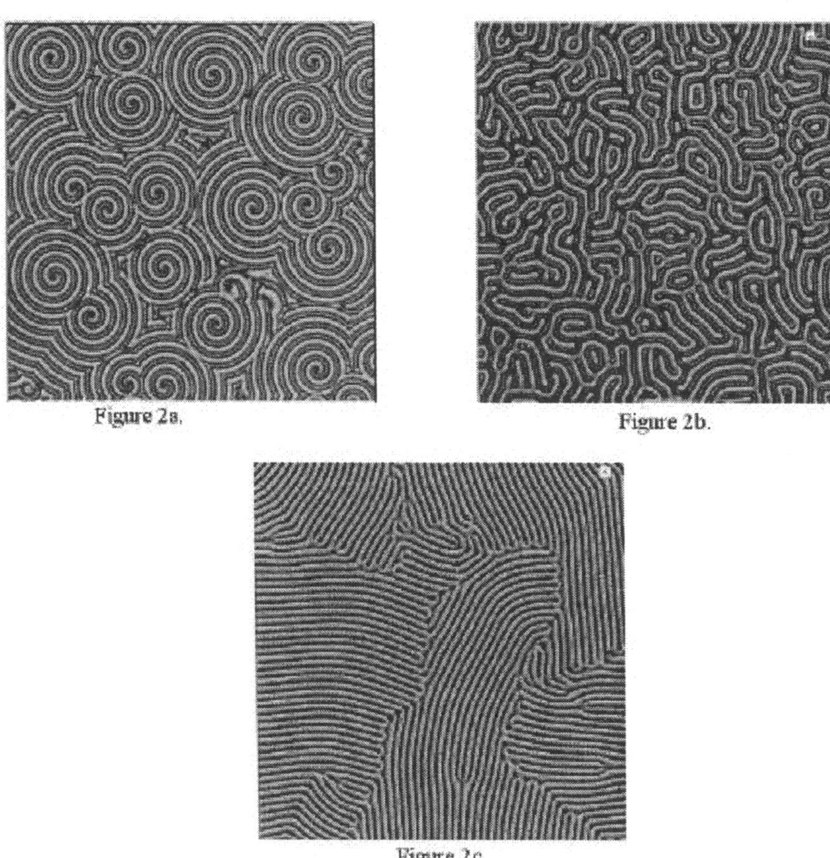

Figure 2: P. Coullet.

2. Vers une neurocognition des formes

Les formes ne sont pas seulement des structures organisées émergeant de la physique interne des substrats. Ce sont aussi des structures perceptives et cognitives construites par le cerveau. Grâce aux résultats spectaculaires des neurosciences, on commence à comprendre comment le cerveau peut élaborer la géométrie des morphologies visuelles. Je me bornerai à un exemple très simple concernant le processus d'intégration des contours à partir de leur représentation distribuée dans le cortex. Comment, alors que la détection de contour est implémentée de façon terriblement locale et distribuée dans le cortex, des formes macro possédant de la cohérence globale peuvent-elles émerger? Cette étonnante performance résulte d'une part de l'architecture fonctionnelle des aires du cortex et d'autre part de phénomènes dits de binding (de liage) assurant la cohérence. Restreignons-nous à la première des aires visuelles, l'aire V1, où débute le traitement des formes. Le signal rétinien y est analysé par des neurones dont les champs récepteurs possèdent *une préférence orientationnelle*. Trois niveaux d'organisation sont particulièrement importants: la rétinotopie, la structure laminaire et la structure colomnaire.

(i)   La rétinotopie signifie qu'il existe des applications de la rétine sur les couches corticales qui préservent la topographie rétinienne. Il existe par exemple une représentation conforme (de type logarithme complexe) entre la rétine et la couche 4C (partie de la couche 4 de V1 où se terminent majoritairement les fibres issues du corps genouillé latéral).

(ii)   La structure laminaire (d'épaisseur environ 1,8 mm) est constituée de 6 couches "horizontales" (i.e. parallèles à la surface du cortex).

(iii) La structure colomnaire et hypercolomnaire est la grande découverte des Prix Nobel Hubel et Wiesel au début des années 60. Il existe dans l'aire V1 des neurones sensibles à l'orientation, à la dominance oculaire et à la couleur. Ce sont les premiers qui nous intéressent ici. Ils détectent des couples (a, p) d'une position rétinienne a et d'une orientation locale p en a. Par des méthodes sophistiquées d'enregistrement de réponse à des stimuli appropriés (barres orientées traversant le champ récepteur des cellules), on a pu montrer que, perpendiculairement à la surface du cortex, la position rétinienne a et l'orientation préférentielle p restent à peu près constantes. Cette redondance "verticale" définit les *colonnes d'orientation*. En revanche, parallèlement à la surface du cortex, la position a varie peu dans des domaines où l'orientation préférentielle p varie au contraire de 0° à 180° par pas d'environ 10° tous les 50-100 µ. Ce regroupement "horizontal" de colonnes définit une *hypercolonne d'orientation* qui est un module neuronal d'environ 500 µ-1 mm.

Des études récentes (années 90) ont montré que ces hypercolonnes s'organisent de façon géométriquement très précise en "roues d'orientation" appelées des *pinwheels*. La couche corticale est réticulée par un réseau de singularités qui sont des centres de pinwheels. Autour de ces points singuliers toutes les orientations sont représentées et ces roues d'orientation locales se raccordent ensuite entre elles. Cette structure est intéressante car elle permet d'implémenter dans des couches corticales de dimension 2 une structure abstraite qui est en fait de dimension 3: 2 degrés de liberté pour les positions a et un degré de liberté pour l'orientation p. (Cf. fig. 3).

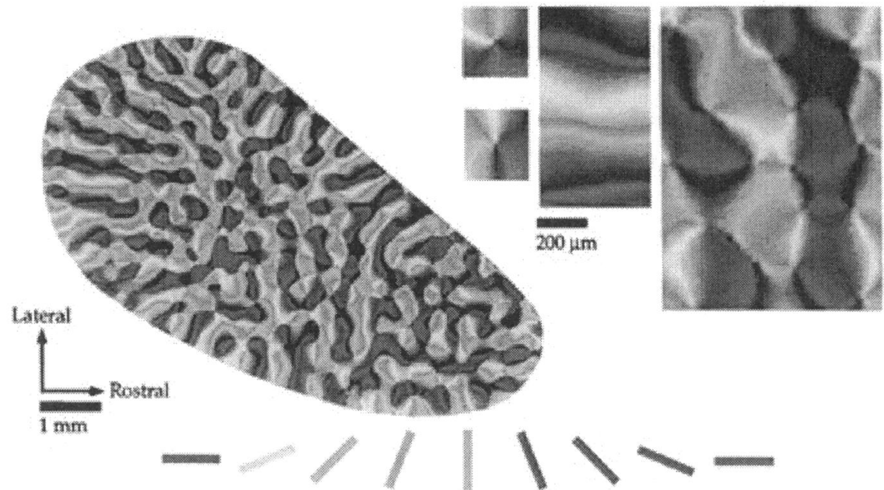

Figure 3: W. Bosking.

À travers cette architecture fonctionnelle, à chaque position a de la rétine R se trouve associé de façon rétinotopique un exemplaire (discrétisé) P_a de l'espace P des directions p du plan. Mathématiquement, une telle structure s'appelle une *fibration* de base R et de fibre P et il existe donc une implémentation neuronale de cette entité géométrique. L'ensemble des "projections" (au sens neurophysiologique) ascendantes des voies rétino-géniculo-corticales implémente la projection (au sens géométrique) $\pi : R \times P \to R$. Mais cette structure purement "verticale" ne suffit pas. Pour qu'il y ait cohérence globale, il faut pouvoir *comparer* entre elles des fibres P_a et P_b rétinotopiquement voisines. Cela se trouve réalisé à travers des connexions "horizontales" cortico-corticales. Deux types de structures fondamentales ont été découvertes.

On a d'abord découvert que des connexions cortico-corticales reliaient des cellules de même orientation dans des hypercolonnes voisines. Cela signifie que le système est capable de savoir, pour b voisin de a, si une orientation en a est la même qu'une orientation en b. Autrement dit, les connexions verticales donnent un sens pour le système aux relations entre (a,p) et (a,q) et les connexions horizontales donnent un sens aux relations entre (a,p) et (b,p).

Mais il y a plus. On a en effet pu montrer que des connexions cortico-corticales à longue portée corrèlent *à longue distance* des cellules codant des couples (a,p) et (b,p) *telles que p soit l'orientation de l'axe ab*. Il s'agit là d'un résultat absolument remarquable de William Bosking montrant que ce que les géomètres appellent la *structure de contact* de la fibration $\pi : R \times P \to R$ est implémenté neurologiquement.

Cette architecture fonctionnelle de V1 permet de comprendre l'intégration des contours, c'est-à-dire la façon dont un grand nombre de couples (a,p) détectés localement peuvent s'agréger en une forme globale cohérente? C'est un problème typique de Gestalt, problème hautement non trivial puisque, comme le montrent les célèbres expériences de Kanizsa, le cortex est capable d'extrapoler des contours virtuels à très longue distance. Dans ces expériences, le phénomène d'émergence se révèle dans toute sa portée *puisque le contour émergent n'existe même pas dans les données sensorielles*.

En fait ces phénomènes sont des cas particuliers très simples de phénomènes de *liage* (de binding). Le binding est un problème cognitif central, celui de la "constituance" et de la "compositionalité" des représentations mentales. C'est le problème classique du tout et des parties, le problème dit *méréologique* des structures. Au niveau neuronal, les représentations mentales sont implémentées de façon *distribuée* sur un très grand nombre d'unités élémentaires. Comment éviter la "catastrophe" dissolvant les structures parties / tout dans le chaos pointilliste de l'implémentation? Comment arriver à extraire des constituants possédant une unité individuée? Comment coder leurs liens relationnels?

L'une des hypothèses actuellement les plus discutées—qui remonte à des travaux de Christoph von der Malsburg, 1981—repose sur le *codage temporel fin* des processus mentaux. Elle est que la cohérence structurale—l'unité—des constituants d'une représentation mentale se trouve encodée dans la dynamique de l'activité neuronale sous-jacente, dans ses corrélations temporelles et, plus précisément, dans la *synchronisation* (accrochage de fréquence et de phase) de réponses neuronales oscillatoires. L'idée est donc que la cohérence temporelle rapide (de l'ordre de la ms) code la cohérence structurale. La phase commune des oscillateurs synchronisés implémentant un constituant peut alors servir de label pour ce constituant dans des processus de traitement ultérieurs. D'où aussi le nom de "labeling hypothesis".

Il existe de très nombreuses confirmations expérimentales d'oscillations synchronisées (dans la bande de fréquence γ des 40 Hz) des colonnes et hypercolonnes corticales, la synchronisation étant sensible à la constituance des stimuli et à la cohérence de leurs constituants (travaux de Eckhorn, Gray, Singer, König, Engel, 1992).

Ces résultats ont été fort débattus et sont en partie controversés. Certains pensent même qu'ils sont partiellement épiphénoménaux. Il faut dire qu'ils sont fort délicats à obtenir et que de nombreux paramètres y interfèrent. Certaines conditions expérimentales spécifiques renforcent peut-être les oscillations. Qui plus est, ils concernent plus l'individuation des constituants que la structure de leurs assemblages. Mais il valident néanmoins une idée directrice. Même si l'on simplifie et idéalise celle-ci outrancièrement, elle conduit, comme cela a été le cas avec les réseaux de neurones formels, à des problèmes mathématiques d'une grande difficulté.

En ce qui concerne la modélisation, on montre d'abord que des colonnes corticales peuvent effectivement fonctionner comme des oscillateurs élémentaires. Elles sont constituées d'un grand nombre de neurones excitateurs et inhibiteurs. En moyennant sur ces deux groupes les équations standard des réseaux de neurones, on obtient un système de deux équations (équations de Wilson-Cowan). On montre alors que l'état d'équilibre subit une bifurcation de Hopf lorsque l'intensité du stimulus dépasse un certain seuil.

Ceci dit, le choix des types d'oscillateurs pose lui même un problème complexe. Il existe en effet au moins 3 types différents d'oscillateurs.

1. les oscillateurs harmoniques et leurs variantes (cycles limites uniformes);

2. les cycles d'hystérésis apparaissant dans les systèmes lents/rapides dont la variété lente a une forme normale cubique (cycles de Van der Pol);

3. les cycles limites avec saut discontinu de type firing (décharge).

D'autre part, le choix des formes de couplage présente également des alternatives. A côté des couplages de type sinus des différences de phases, il y a des modèles (Mirollo, Strogatz, Kuramoto, 1991) où l'on couple les oscillateurs par des pulses. D'autres modèles de synchronisation ont été proposés, comme les "synfire chains" d'Abeles et Bienenstock (ondes de synchronisation le long de chaînes séquentielles).

Quoi qu''il en soit, la simplification maximale conduit à étudier des réseaux constitués d'un grand nombre N d'oscillateurs F_i dont la fréquence propre ω_i dépend de l'intensité du stimulus à la position i. Soient θ_i leurs phases et $\varphi_i = \theta_{i+1} - \theta_i$ leurs différences de phases. Les équations du système sont du type:

$$\dot{\theta}_i = \omega_i - H(\varphi_1, \ldots, \varphi_{N-1}).$$

Les systèmes les plus courants sont du type:

$$\dot{\theta}_i = \omega_i - \sum_{j=1}^{j=N} K_{ij}\, \sin(\theta_i - \theta_j)$$

où les K_{ij} sont des constantes de couplage. Ce sont des systèmes typiquement complexes que l'on peut étudier avec des méthodes de physique statistique (travaux de Kuramoto, Daido, etc.) et de dynamique qualitative (travaux d'Ermentrout et Kopell, etc.).

Dans le cas d'une seule constante de couplage et d'une totale connectivité, Kuramoto 1987 a analysé en détail le système:

$$\dot{\theta}_i = \omega_i - \frac{K}{N} \sum_{j=1}^{j=N} \sin(\theta_i - \theta_j)$$

Pour ce faire, il a introduit le *paramètre d'ordre* qu'est la phase moyenne :

$$Z(t) = |Z(t)| e^{i\theta_0(t)} = \frac{1}{N} \sum_{j=1}^{j=N} e^{i\theta_j(t)}$$

et a étudié le système équivalent:

$$\dot{\theta}_i = \omega_i - K|Z| \sum_{j=1}^{j=N} \sin(\theta_i - \theta_0)$$

Si les fréquences ω_i sont tirées au hasard suivant une loi $g(\omega)$ représentant les régularités statistiques de l'environnement (en prenant un repère tournant on peut supposer g centrée sur 0), la synchronisation globale est une *transition de phase* s'effectuant pour la valeur critique $K_c = 2/\pi g(0)$ de la constante de couplage.

Pour le montrer, Kuramoto cherche d'abord des solutions $Z =$ constante. Après avoir classé les oscillateurs en deux groupes:

(i) le S-groupe des oscillateurs pouvant se synchroniser i.e. satisfaisant

$$\dot{\theta}_i = 0 \text{ et donc } \left| \frac{\omega_i}{KZ} \right| \leq 1$$

(ii) le D-groupe des oscillateurs ne le pouvant pas parce que

$$\left|\frac{\omega_i}{KZ}\right| > 1$$

il montre que seul le S-groupe intervient dans la synchronisation.

En écrivant que

$$Z = \int_0^{2\pi} n_0(\theta,t)e^{i\theta}d\theta$$

où $n_0(\theta,t)$ est la distribution des phases à l'équilibre au temps t et en écrivant que

$$n_0(\theta,t)d\theta = g(\omega)d\omega \text{ avec } \omega = K|Z|\sin(\theta - \theta_0)$$

il obtient une équation d'auto-consistance $Z = S(Z)$ qu'il développe au voisinage de $Z = 0$. D'où l'équation (qui est une forme normale):

$$\varepsilon Z - \beta|Z|^2 Z = 0$$

$$\text{avec } \varepsilon = \frac{K - K_c}{K_c}, \beta = -\frac{\pi}{16}K_c^3 g''(0).$$

L'analyse de la stabilité des solutions montre que la solution $Z=0$, qui est stable pour $K\approx 0$ (oscillateurs découplés), devient instable à la traversée de $Z=Z_c$.

Kuramoto établit ensuite, sous une hypothèse de quasi-adiabaticité, l'évolution (lente) du paramètre d'ordre Z. Elle est régie par une équation du type:

$$\xi\frac{dZ}{dt}|KZ|^{-1} = \varepsilon Z - \beta|Z|^2 \overset{.}{Z}$$

Il étudie ensuite les *fluctuations*, en particulier au voisinage du point critique lorsqu'elles deviennent géantes et entraînent la transition de phase.

Ces résultats montrent que la synchronisation est un phénomène typique d'organisation collective émergente.

Daido 1990 a étudié quant à lui les systèmes:

$$\overset{.}{\theta_i} = \omega_i - K\sum_{j\in V_i}\sin(\theta_i - \theta_j)$$

où V_i est l'ensemble des plus proches voisins de l'oscillateur de rang i sur un réseau cubique de dimension d. Par des méthodes du *groupe de renormalisation*, il a montré que, si la loi $g(\omega)$ est asymptotiquement une loi de puissance:

$$g(\omega) \approx |\omega|^{-\alpha-1}, \alpha \in \,]0,2[$$

alors le système est équivalent à un système découplé (et donc non synchronisable) pour:

$$\beta = 1 - \frac{1}{\alpha} - \frac{1}{d} < 0$$

De façon plus précise, si l'on décompose le réseau en $M = L^d$ blocs de taille L (l'unité de longueur étant la maille du réseau initial) et si l'on moyenne les fréquences ω_i et les phases θ_i sur les blocs B_k (ce qui donne des fréquences $?_k$ et des phases φ_k), on obtient les équations suivantes:

$$\dot{\varphi}_k = \Omega_k - \frac{K}{M} \sum_{\substack{i \in B_k \\ j \in B_l}} \sin\left(\varphi_l - \varphi_k + \psi_{i,j} - \psi_{k,j}\right)$$

avec $\theta_j = \varphi_m + \psi_{m,j}$ si $j \in B_m$.

L'opération de renormalisation est alors donnée par:

$$\begin{cases} \tau = tM^{(1-\alpha)/\alpha} \\ \varphi_k^* = \varphi_k - \dfrac{\gamma_M}{M} t \end{cases}$$

où γ_M est défini par le fait que la fréquence

$$\omega_n^* = \frac{\sum_{i=1}^{i=n} \omega_i - \gamma_n}{n^{1/\alpha}}$$

obéit à une loi de distribution stable de fonction caractéristique

$$\left\langle \exp\left(iz\omega^*\right)\right\rangle = \exp\left(-|z|^\alpha\right)$$

Rappelons que si X est une variable aléatoire de fonction de répartition

$$F(x) = p(X < x)$$

sa fonction caractéristique est la transformée de Fourier

$$G(z) = \left\langle \exp\left(izx\right)\right\rangle = \int \exp(izx)dF(x)$$

Les lois *stables* sont des lois indéfiniment divisibles (c'est-à-dire qui peuvent être considérées comme des sommes de variables aléatoires infiniments petites indépendantes: toutes les $(G(z))^\alpha$ pour $\alpha > 0$ sont des fonctions caractéristiques) dont la classe est stable par combinaisons linéaires. On montre que leur fonction caractéristique est alors du type

$$G(z) = \exp\left[\left(-c_0 + \frac{iz}{|z|} c_1\right)|z|^\alpha\right]$$

avec $\alpha \in\ ]0,2[$, $c_0 \geq 0$ et $\left| c_1 \cos(\pi\alpha\,/\,2) \right| < \left| c_0 \sin(\pi\alpha\,/\,2) \right|$. On a ici $c_0 = 1$ et $c_1 = 0$.

Daido obtient ainsi les équations renormalisées:

$$\frac{d\varphi_k^*}{d\tau} = \omega_{M,k}^* - KM^\beta \sum_{l \in J_k} \sin{}_{lk}^* \left(\varphi_l^* - \varphi_k^* \right)$$

avec $\beta = 1\text{-}1/\alpha\text{-}1/d$, $J_k = \{$blocs B_l voisins du bloc $B_k\}$, le couplage effectif étant donné par:

$$\sin{}_{lk}^*(\varphi) = M^{(1-d)/d} \sum_{\substack{i \in B_l \\ j \in B_k}} \sin\left(\varphi + \psi_{l,i} - \psi_{k,j} \right)$$

où pour $j \in B_m$, $\psi_{m,j}$ est l'écart de la phase θ_j à la phase moyenne φ_m sur B_m : $\theta_j = \varphi_m + \psi_{m,j}$.

Le fait de base est que pour $\beta < 0$, le système est attiré par le point fixe trivial

$$\frac{d\varphi_k^*}{d\tau} = \omega_k^* \text{ où } \omega_k^* = \lim_{M \to \infty} \omega_{M,k}^*$$

du groupe de renormalisation. Les interactions tendent vers 0 et il ne peut y avoir de synchronisation. Cela n'empêche évidemment pas un phénomène de *clustering*. Au fur et à mesure que le couplage K augmente, des oscillateurs de plus en plus nombreux se synchronisent. Mais il ne s'agit plus d'une transition de phase.

Quant à Erik Lumer, il a étudié à la suite d'Huberman des systèmes

$$\dot{\theta}_i = \omega_i - \sum_{j=1}^{j=N} K_{ij} \sin(\theta_i - \theta_j)$$

où $K_{ij} = Kd(l_{ij})$, l_{ij} étant la distance ultramétrique sur un arbre exprimant une hiérarchie de constantes de couplage. Si b est le degré de branchement de l'arbre et L sa hauteur, on impose la condition:

$$\sum_{l=1}^{l=L} (b-1)b^{l-1}d(l) \underset{L \to \infty}{\to} 1$$

($(b-1)b^{l-1}$ étant le nombre des oscillateurs à distance exactement l d'un oscillateur donné). On peut prendre par exemple $d(l) = \dfrac{1}{(b-1)b^{l-1}} \cdot \dfrac{a-1}{a^l}$.

On montre alors, toujours au moyen des méthodes du groupe de renormalisation, qu'il n'existe de synchronisation par transition de phase que si $b \geq a^{\alpha/(\alpha-1)}$ ($\alpha \in\ ]1,2]$). Sinon il existe une synchronisation progressive, par paliers.

Une fois mieux comprises les propriétés de synchronisation de tels systèmes d'oscillateurs, on peut construire, sur la base de la "labeling hypothesis", des modèles de fonctions cognitives de haut niveau. Par exemple Lumer 1992 a proposé une théorie du processus d'*attention* permettant de se focaliser sur un constituant d'une scène perceptive. Elle consiste à extraire la phase d'un groupe synchronisé au moyen d'un "phase tracker" et à l'utiliser comme label.

Bibliographie

Amit D.: 1989, *Modeling Brain Function*. Cambridge University Press.

Atiya A., Baldi P.: 1989, "Oscillations and Synchronisation in Neural Networks: an Exploration of the Labeling Hypothesis", *International Journal of Neural Systems* 1, 2, pp. 103-124.

Bonhoeffer T., Grinvald A.: 1991, "Iso-orientation domains in cat visual cortex are arranged in pinwheel-like patterns", *Nature* vol. 353, pp. 429-431.

Bosking W. et al.: 1997, "Orientation Selectivity and the Arrangement of Horizontal Connections in the Tree Shrew Striate Cortex", *The Journal of Neuroscience* 17, 6, pp. 2112-2127.

Coullet P., Emilsson K.: 1992, "Strong resonances of spatially distributed oscillators : a laboratory to study patterns and defects", *Physica D* 61, pp. 119-131.

Daido H.: 1990, "Intrinsic Fluctuations and a Phase Transition in a Class of Large Populations of Interacting Oscillators", *Journal of Statistical Physics* 60, 5/6, pp. 753-800.

De Kepper P. et al.: 1998, "Taches, rayures et labyrinthes", *La Recherche* 305, pp. 84-89.

Engel A., König P., Gray C., Singer W.: 1992, "Temporal Coding by Coherent Oscillations as a Potential Solution to the Binding Problem: Physiological Evidence", in H. Schuster (ed.): *Non Linear Dynamics and Neural Networks*. Berlin: Springer.

Field D.J., Hayes A., Hess R.F.: 1993, "Contour integration by the human visual system", *Vision Research* vol. 33, 2, pp. 173-193.

Gray C.M., König P., Engel A.K., Singer W.: 1989, "Oscillatory responses in cat visual cortex exhibit inter-columnar synchronization which reflects global stimulus properties", *Nature* vol. 338, pp. 334-337.

Kanizsa G.: 1979, *Organization in Vision : Essays on Visual Perception*. Praeger. Trad. française *La Grammaire du Voir*. Diderot Editeur, Paris.

Kopell N., Ermentrout G.B.: 1990, "Phase Transitions and Other Phenomena in Chains of Coupled Oscillators", *SIAM J. Appl. Math.* 50, 4, pp. 1014-1052.

Kuramoto Y., Nishikawa I.: 1987, "Statistical Macrodynamics of Large Dynamical Systems. Case of a Phase Transition in Oscillator Communities", *Journal of Statistical Physics* 49, 3/4, pp. 569-605.

Lumer E. D., Huberman B. A.: 1992, "Binding Hierarchies: A Basis for Dynamic Perceptual Grouping", *Neural Computation* 4, pp. 341-355.

Marr D.: 1982, *Vision*. San Francisco: Freeman.

Meinhardt H.: 1995, *The Algorithmic Beauty of Seashells*. Springer.

Petitot J.: 1990, "Le Physique, le Morphologique, le Symbolique. Remarques sur la Vision", *Revue de Synthèse* 1-2, pp. 139-183.

— 1992, *Physique du Sens*. Paris: Editions du CNRS.

— 1994, "Morphodynamics and Attractor Syntax", in T. von Gelder, B. Port (eds.): *Mind as Motion*. MIT Press.

— 1999, "Morphological Eidetics for Phenomenology of Perception", in J. Petitot, F. J. Varela, J.-M. Roy, B. Pachoud (eds.): *Naturalizing Phenomenology: Issues in Contemporary Phenomenology and Cognitive Science*. Stanford: Stanford University Press, pp. 330-371.

Petitot J., Tondut Y.: 1999, *Vers une Neuro-géométrie. Fibrations corticales, structures de contact et contours subjectifs modaux*, Numéro spécial de *Mathématiques, Informatique et Sciences Humaines* 145 (EHESS, Paris), pp. 5-101.

Petry S., Meyer G.E. (eds): 1987, *The Perception of Illusory Contours*. Springer.

Thom R.: 1972, *Stabilité structurelle et Morphogenèse*. New York : Benjamin. Paris: Ediscience.

Part II

Complexity and Emergence in Natural Science

7.   Emergence in Physics: the Case of Classical Physics

Roland Omnès

Lab. of Theoretical Physics, University of Paris Sud, France

Physics is a most convenient place for studying emergence, as I intend to show here. The idea of emergence will first be considered in rather general terms before concentrating on a special topic: the emergence of classical physics from quantum theory. The answer to the question of emergence has seen much progress in the last two decades, and it is particularly instructive from a philosophical standpoint, which is the one we are of course most interested in.

Physics is a very mature science. A long process of discovery has led to our seeing it as resting on two basic pillars of knowledge: a container, which is space-time, and a content, consisting ultimately of elementary particles. Our understanding of space-time and gravitation is based on the general theory of relativity, whereas the description of particles is now given by the standard model of quarks and leptons and rests on quantum mechanics. Much work is also presently devoted to a search for a deeper synthesis of these two faces of reality, but it will not be considered here, since the philosophy of science should be primarily concerned with a body of knowledge that has met every attempt at falsification.

Most physicists agree that the best way to understand their field of science consists in considering it as built on some deep fundamental principles (although there are still two different kinds of principles when either space-time or particles are concerned). The multitudinous aspects of physical reality are then conveniently classified according to the type of objects and phenomena that are met in practice and traditionally covered by particular disciplines such as nuclear physics, atomic physics, solid state physics, electromagnetism, optics, thermodynamics, chemistry and so on.

Every such subfield of physics is considered as emerging from the basic principles, in a way that we will of course have to explain.

Before that, one must recognize a practical difficulty in this vision of physics (which is obviously strongly structuralist and therefore somewhat reminiscent of the Hilbert-Bourbaki conception of mathematics). Rather than building on the structures of set theory, algebra, topology and so on, one builds on space-time and quanta in their most refined expression. But whereas the first elements of set theory, algebra and such are easily grasped by a beginner, the amount of knowledge of quantum field theory that would permit linking it to, say, chemistry or fluid mechanics, is frightful. We assume that, as a matter of fact, there exists no book which makes completely explicit the structure of physics, no Bourbaki "elements of physics". The nearest to it is a beautiful series of treatises by Landau and Lifschitz, but they had to contend with the realities of pedagogy by beginning with the end: classical mechanics[1].

It should be no surprise in these conditions that some hard-boiled skeptics criticize the structuralist conception of physics as an illusion, if not a fraud[2]. They certainly have a point. When looking more carefully at some of the supposed links between different fields of physics, one must acknowledge that there is yet no complete derivation of nuclear forces from the quark structure of nucleons, nor of the existence of crystals from the principles of quantum mechanics, nor of quite a few similar transitions we would like to call "emergences". Nancy Cartwright has pointed out some of these weaknesses in physics and other sciences so that, in her opinion, science itself has no sign of unity and should be better called a "patchwork". Her critique shows by the way that there is a strong connection between the notion of emergence and the question of the unity of science, or even the internal unity of one science (such as physics for instance).

1. What is emergence? Are there emergences?

The idea of emergence can also be related to reductionism. To say for instance that the main features of biological phenomena emerge from the laws of chemistry and physics is more or less a politically correct way of

---

[1] See Landau and Lifschitz 1959.
[2] See Cartwright 1983.

reducing biology to physics (after reducing primarily chemistry to physics). One must obviously avoid entering into such messy marshes and, if only for that reason, the present discussion will be strictly restricted to physics.

I propose then to define emergence as follows: A field of science B emerges from another field of science A ($A{\rightarrow}B$) when the concepts necessary for a formulation of B as well as the laws specific to B can be derived from the concepts and laws belonging to A.

I do not know how much of this definition might be extended outside physics, because this science has two important peculiarities: (a) Every field in it is traditionally constructed so that its specific domain, its constitutive concepts, and its specific laws are explicitly stated, at least as far as possible. (b) Physics is a formal science. By "formal", I mean a science (or a field of science) in which the constitutive concepts are well defined as are the basic laws and, furthermore, both can be completely expressed in the language of mathematics. This is true for instance of classical mechanics, since its formulation by Newton. In the case of quantum mechanics, the formal character is still much stronger since not only *can* its concepts and laws be expressed mathematically, but they cannot be fully and usefully expressed otherwise (for a discussion of formal sciences, see Omnès 1994a).

These two characteristics of physics and its specialized fields insure that our definition of emergence makes sense in their framework. The concepts and laws we mentioned can be read out from a standard treatise about the two fields of science A and B; and the "derivation" of B from A can be supposed to be a mathematical construction. This is at least how one can give a meaning to emergence by relying on theory.

If this is agreed, there is no doubt that emergences are many in physics, and some examples may allow us to refine the notion of emergence itself. Take for instance the emergence of optics from electromagnetism, as predicted by Maxwell and checked by Hertz. This is a very simple case since the "derivation" turns out simply to make explicit some concepts of optics (intensity, polarization, etc.) as functions of electric and magnetic fields.

The emergence of equilibrium thermodynamics from statistical mechanics is almost as easy, since the basic thermodynamical parameters such as temperature, free energy, and entropy, are necessary mathematical ingredients of a statistical distribution. Thermodynamics out of equilibrium has on the contrary not yet quite emerged from particle physics since it involves the existence of a specific direction of time, which does not appear in the quantum axioms. (It may however be beginning to emerge, as shown for instance in Omnès 1999)

There are cases when emergence brings out a much deeper understanding of a particular subject. Ferromagnetism for instance is explained as emerging from atomic physics (through quantum mechanics) as an *electric* interaction (hence its strength), while its characteristic effect of magnetization is a macroscopic manifestation of an enforced parallelism between some electron spins, which itself results from the exclusion principle.

Another emergence is very interesting from the standpoint of epistemology. You may remember a famous saying by Hertz concerning the meaning of physical principles in the case of Maxwell's laws: "There is no other principle on which these laws are founded than Maxwell's *equations* themselves". In 1940, Eugene Wigner (later with Valentine Bargmann) investigated the mathematical consequences of the principle of special relativity in the framework of quantum mechanics. He found that this principle implies the existence of a mass and a spin for every particle. If a particle has zero mass and spin 1 and if many such particles are produced together, some of the quantum observables characterizing them include an electric and a magnetic field, which obey Maxwell's equations in a vacuum. So, there was a very simple principle behind these equations after all: photons exist, with zero mass and spin 1!

2. The case of classical physics

Soon after the discovery of the basic laws of quantum mechanics in the years 1925-1926, it was realized that quanta rule in a world very different from the one we are used to. Already at the end of 1926, Max Born had shown that the wave function represents probabilities, so that to accept the existence of wave functions implies an acceptance of a non-deterministic atomic world. The principle of causality, so basic in philosophy, was therefore rejected by the new physics. Soon after, in 1927, Heisenberg discovered his famous inequalities linking the uncertainties regarding position and momentum. An immediate consequence was the impossibility of assuming that particles have a trajectory in space, which implied that one had to give up Leibniz' locality principle and Kant's *a priori* synthetic judgments on space and time. Then, in 1935, Schrödinger spelled out a quantum property that had been noticed earlier by Von Neumann. With the famous example of a cat entering as a part of a measuring apparatus, he claimed a general existence of macroscopic interferences in quantum mechanics, a most strange phenomenon of a cat dead *and* alive that is in fact never observed. Von Neumann and Birkhoff claimed afterwards that this example and others indicate an incompatibility of the world of quantum mechanics with standard logic.

Nancy Cartwright, whom I cited earlier, had a wonderful opportunity for pointing out a major gap, a rent in the patchwork of physics keeping quantum and classical physics apart and out of touch with one another. She had only to quote the highest authorities in physics: Schrödinger and Von Neumann of course, but also Bohr himself. The relation between quantum and classical physics was only evoked by Bohr as some sort of "correspondence" that was never very explicit in spite of a vague "correspondence principle". Heisenberg spoke of an arbitrarily moving frontier separating the two kinds of physics.

Bohr made the strongest argument against any suggestion of the emergence of one kind of physics out of the other. Pointing out that the uniqueness and definiteness of a fact is the fundamental criterion of truth in the natural sciences, he asserted that the language of classical physics is the only logical one we can use with certainty and, as a consequence, the range of truth must be restricted to the phenomena pertaining to classicality. By this *fiat*, quantum and classical physics were separated in essence and made mutually irreducible.

This is a very brief summary of a question which has been much discussed in many books; but, in the present case, it shows that: if there were some sort of emergence between these two *A* and *B*, quantum and classical physics, it could be considered the paradigm of emergence. I wish now to explain how, in the last two decades, the conditional "if there were ... could" has been replaced by "because there is ... should", thus renewing and completely clarifying the notion of emergence.

3. Classical physics emerging

If I were to present a complete discussion of this topic here, I would have to cover the content of at least one or two published books[3], more books to come by other writers, and many important articles by various authors over a period of more than seventy years. The emergence of classical physics from quantum physics is an intrinsic part of a new interpretation of quantum mechanics, which relies on the discovery of new effects or new notions whose names are now becoming more or less familiar: decoherence, semi-classical physics, consistent histories and standardized quantum logic. As one may expect in a period of discovery, not all the authors contributing to these works agree on the exact place and role of the main ideas, and the critics are many. The fact however that some of the critics begin to speak of this approach as "the new dogma" implies that it is probably more than a temporary fashion. More significantly, its agreement with all the known experimental data, including the new ones purporting to check it, gives it much weight.

Unfortunately it will be impossible to completely justify here how classical physics emerges from the quantum principles. It will also not be possible to explain the ideas and techniques that must be used to obtain the results, and I will try rather to stress their meaning. Concerning the controversies about the whole "new interpretation", I wish only to mention that our colleague Bernard d'Espagnat, with his usual discernment, considers this emergence as one of the least controversial results of the new approach[4].

Let us then list the main characteristics of classical physics where they seem to conflict with quantum physics:

1. Some statements relevant to classical physics involve position (when saying for instance where an object is located at some time) or they involve momentum; but many statements involve a simultaneous consideration of position and momentum (when giving for instance the initial conditions of a motion). This is to be contrasted with the non-commutativity of position and momentum in quantum mechanics.
2. Two statements involving clearly distinct positions and momenta (i.e. two clearly different situations in phase space) are mutually exclusive. This means that they cannot hold at the same time. They may however

---

[3] See Omnès 1994 and 1999.
[4] See d'Espagnat 1995.

enter in probabilistic considerations as two mutually exclusive events having definite probabilities which satisfy the usual rules of the probability calculus. This is of course contrasted with quantum interferences (Schrödinger's cat).

3. There is nothing resembling complementarity in classical physics. This means that the various classical properties of a system (at the same or different times) can be considered as a field of propositions in which standard logic holds.

4. Of particular interest in this logic of classical physics is the logical equivalence of two propositions holding at two different times when they are related by determinism. This rather abstract way of looking at determinism is clearer in an example where friction is assumed to be negligible. Let a be a proposition stating the position and velocity of the moon at some time t, and b another proposition of the same kind for a later time t'. When the two statements agree with the predictions of classical dynamics, determinism means that a implies b (prediction). In the absence of friction, b also implies a (retrodiction). Since a and b imply each other, they are logically equivalent. The (apparent) contrast between classical determinism and quantum probabilism is of course well known.

5. When friction is not negligible, time has a definite direction. There is no such direction at the level of quantum laws.

6. Although classical physics can be formulated abstractly in a convenient mathematical framework (configuration space and phase space), it can also be expressed as taking place in ordinary three-dimensional space. Conversely, a wave function describing several particles is defined on their n-dimensional configuration space because of entanglement.

7. Macroscopic objects behave classically (whereas particles and atoms are governed by quantum mechanics).

It turns out that most features of this list can now be derived from the basic axioms of quantum mechanics and, for those that are not yet rigorously derived, interesting new vistas are opening. There are also explicit limits to the range of the validity of this emergence, which are most easily understood as originating from quantum fluctuations. More precisely, some of the results implying emergence are as follows:

1. Let us first define the classical properties we are interested in: In view of the uncertainty relations, a meaningful classical property cannot precisely express position and momentum together. It will only state for instance that a position coordinate lies in an interval $[x—?x, x + ?x]$ and a momentum coordinate in an interval $[p—?p, p + ?p]$, the product $?x.?p$ being much larger than the Planck constant h. Such a classical property should be contrasted with a quantum property, which involves no pairs of non-commuting observables. A quantum property would only state for instance that a position coordinate lies in an interval $[x—?x, x + ?x]$ with no commitment regarding the value of momentum. Von Neumann discovered long ago a beautiful "translation" of quantum properties into the mathematical language of quantum theory[5]. Every quantum property can be associated with a "projection operator" in Hilbert space. This operator contains exactly the same information as the statement of the property. Furthermore, as an observable, it can only take the values 1 (for "true") or 0 (for "false"), as is to be expected of a proposition. But despite his attempts, Von Neumann was unable to extend this idea to classical properties. The answer is a matter of mathematics, and relies on microlocal analysis (the so-called "analysis of the 1970's", which is now the recognized basis of the theory of linear partial differential equations and the most powerful approach to semi-classical physics)[6]. It will suffice for our purpose to say that in place of a unique projection operator expressing a quantum property, a set of "equivalent" projections can express a classical property. The real importance of this result is of course the existence of a common (mathematical) language for expressing classical as well as quantum properties.

2. The fact that classical properties obey standard logic with no hint of quantum complementarity is an apparently obvious statement, which suddenly becomes almost incredible when it is supposed to emerge from quantum theory. It is still true however. To make a long story short, let us say that the proof of this "fact" is a direct outcome of the method of "consistent histories"[7] (the main point of this method being precisely the return of standard logic in the interpretation of quantum mechanics). A more philosophical way of stating this result may refer

---

[5] See Von Neumann 1955.
[6] See Hörmander 1985.
[7] See Omnès 1994, 1994a and 1999, and Griffiths 1984.

to the *category* of reality, as used by Kant, i.e. as the existence of a quality that can be attributed to a proposition when it expresses something real. One may thus say that classical properties are indeed able to express reality.

3. The relation of logical inference between two properties occurring at different times is based on dynamics. In the case of the two classical properties a and b of the moon we mentioned earlier, a microlocal theorem by Yuri Egorov[8] entails that a and b imply each other in two ways: They are logically equivalent from the standpoint of quantum logic, when each of them is described by its own set of quantum projections, and dynamics is governed by the Schrödinger equation. They are also equivalent from the standpoint of classical logic when they refer directly to some regions in phase space and the classical equations of motion are operative. This means that *classical determinism is a direct consequence of quantum mechanics*. This basic result is sometimes misunderstood as meaning that something might be proved from its contrary: determinism from pure probabilism. That would of course be absurd. The real meaning of this theorem remains probabilistic, and its precise statement is that both implications from a to b and from b to a have a probability of error that is explicitly known, and completely negligible in ordinary circumstances.

4. As a matter of fact, the probability of error in determinism becomes comparable to 1 when one tries to descend from the macroscopic to an atomic scale. Then, only quantum physics is applicable. The Egorov theorem also shows exceptional cases where the correspondence between classical and quantum physics does not hold. The most remarkable is the case of a chaotic system when chaos reaches the level of Planck's constant.

5. Another deep result is the removal of macroscopic interferences at a macroscopic scale. It is often expressed by referring to the example of Schrödinger's cat, and states that the two events (of a cat being dead or alive) are exclusive. Each possibility has its own probability that obeys the standard probability calculus. The origin of this result is the "decoherence" effect, which was suggested by Hans-Dieter Zeh in 1970, and has since been much investigated theoretically, and

---

[8] See Hörmander 1985.

observed experimentally in 1996[9]. Unfortunately I cannot enter into its analysis here. Let me therefore only mention that it is due to an extreme sensitivity of the (local) phase of a many-particle wave function to the gross features representing its macroscopic state. The removal of macroscopic interferences is then a dynamical effect resulting from destructive interferences (loss of phase coherence) at a microscopic level. Decoherence is by far the most efficient quantum effect with observable consequences on a large scale.

6.   In order to avoid some misunderstanding, one must stress that the results on decoherence, like those on determinism, have exceptions. Some clever superconducting devices show little or no decoherence, and behave quantum-mechanically although they are macroscopic. As first noted by Anthony Leggett[10], one thus cannot consider the old axiom "macroscopic = classical" as being universal.

7.   Among the classical features we mentioned earlier, there were the account of friction (with its arrow of time) and the possibility of describing classical physics "à la Newton" in ordinary space rather than "à la Lagrange" in phase space. The situation regarding these two points is still much less demonstrative than it is for the previous results, although the approach through "decoherence + semi-classical analysis + consistent histories" sheds much light on it. Since, however, hints and insights cannot be compared with theoretical proofs and experimental checks, I prefer not to enter into this moving field and refer rather to a more detailed account[11].

4. A paradigm of emergence

The derivation of classical physics from the basic rules of quantum mechanics is certainly worth considering as a paradigm of emergence. Like most paradigms, it suggests many possible extensions and comments, either plain or bold, of which I will briefly mention only a few.

No problem of emergence has kept scientists so active for so long, except of course for the problems arising from biology, which are much

---

[9] See Brune et al. 1996.

[10] See Leggett 1980.

[11] See Omnès 1999.

more fuzzy. The reason for this insistence was the importance of the problem, which we have already indicated, and a refusal to have two different kinds of laws—two apparently incompatible ways of thinking—in one and the same basic science. Conversely, no other emergence seemed to confront such formidable obstacles and apparent contradictions, such as determinism vs. probabilism, the superposition principle and its intrinsic interferences vs. the sharpness of facts, and even doubts about the validity of logic.

Conversely, the other problems of emergence or the knowledge gaps that still exist in physics are much less impressive. The emergence of nuclear physics from quark physics for instance shows no *a priori* contradiction. It seems only to stand as a difficult theoretical problem on which progress is slowly being made. The emergence of superconductivity from solid state physics, to cite an older example, was perhaps a more difficult problem than this, but it was solved forty years ago.

The critique by Cartwright and others of the unity of science, or at least the unity of physics, must be judged with these examples in mind. Some gaps indeed exist, but they are of a different sort. The most striking one, which was always put forward as the paradigm, was precisely the chasm between classical and quantum physics, since the foundations of physics were in question; but it has now been practically bridged. There was also a gap between solid state physics and superconductivity, but superconductivity is such an important effect that no effort was spared in trying to understand it, and it has been solved. Other gaps, between particle physics and atomic physics, atomic physics and solid state physics, quarks and nuclei, etc. could still be invoked by skeptical philosophers; but here the sociology of science has its word to say. Suppose that a physicist is willing to attack one of these problems for the sake of philosophical consistency. What will happen? Most probably, she will not obtain much of a result, and her career will suffer. Or she will succeed after much effort. What will then be the reaction of the physics community? Most probably something like: "Well done, but so what? We were sure of the result in advance".

Criticism is easy, but art (or science) is difficult. I must say that a sweeping general statement like Cartwright's against the consistency of science is to my mind much less impressive than the accumulation of the past gaps that have now been bridged, and the persistent trend of progress toward unity. The search for some new emergences and convergences remains a reasonable and fecund program in view of this past history, and one may expect the next emergence to be the irreversibility emerging from the reversible basic laws, and the next convergence to be between the theories of space-time and of particles.

My last word is on the meaning of emergence, at least in the case being considered here. We have seen how the basic concepts and laws of science *B* (classical physics) could be derived from those of science *A* (quantum physics). But this is not reductionism. Reductionism would conclude that the concepts and laws of *A* supersede and replace those of *B*. Emergence shows on the contrary that the concepts of classical physics are much more efficient, clear and well-suited to their object than the form they would take in the framework of *A*. It would be ill-considered to think of a table via a wave function, or to spell out a family of projections in Hilbert space rather than saying "my pencil is falling from the table". Emergence is not only a deep sign of the consistency of knowledge. It is essentially creative.

References

Brune, M., Hagley, E., Dreyer, J., Maître, X., Maali, A., Wunderlich, C., Raimond, J.-M., Haroche, S.: 1996, *Phys; Rev. Lett.,* 77, 4887.

Cartwright, N.: 1983, *How the Laws of Physics Lie*. London: Routledge.

d'Espagnat, B.: 1995, *Veiled Reality*. Reading: Addison-Wesley.

Griffiths, R. G.: 1984, *J. Stat. Phys.*, 36, 219.

Hörmander, L.: 1985, *The Analysis of Partial Differential Operators*, 3 vol. Berlin: Springer.

Landau, D. and Lifschitz E. M.: 1959, *Course of Theoretical Physics*, 9 volumes. London: Pergamon Press.

Leggett, A. J.: 1980, *Progr. Theor. Phys., Supplement,* 69, 1.

Omnès, R.: 1994, *Interpretation of Quantum Mechanics*. Princeton University Press.

— 1994a, *Philosophie de la science contemporaine*. Paris: Gallimard. In English, translation by A. Sangalli: *Quantum Philosophy*, Princeton University Press, 1999.

— 1999, *Understanding Quantum Mechanics*. Princeton University Press.

Von Neumann, J.: 1955, *Mathematical Foundations of Quantum Mechanics*. Princeton University Press.

8.   Classical Properties in a Quantum-Mechanical World

Alberto Cordero

City University of New York, Queens College & The Graduate Center

Summary

One major project in the foundations of physics concerns the insertion of classical physics into the framework of quantum theory. However, attempts in that direction are routinely accused of getting classical behavior into the picture only at the most superficial, instrumental level. In this paper I examine the strengths and weaknesses of this general reaction against the unification project. My focal point is a widely agreed list of desiderata for what a satisfactory account should manage to claim about the emergence of classical properties and behaviors in a quantum mechanical world. I consider one particular but typical unification proposal, the so-called "GRW approach", and evaluate it against the mentioned desiderata. Some of the shortcomings that come to view are specific to the GRW program, but some are interestingly general, applying in fact to all the best unification programs today. A critical examination of extant charges against the best current unification projects then follows, leading to the identification of some disputable presuppositions about the status of Boolean structures in physics. The paper closes with an argument for relaxing such presuppositions, along with a brief discussion of the conceptual possibilities that open up when the original list of desiderata is unfettered from the traditional Boolean conception of classical properties.

1. Classical systems in standard quantum theory

Many physical objects behave as classical mechanical says they should: their property state is both sharp and complete at all times; their mechanical evolution proceeds deterministically; and their interactions weaken with distance in all respects, which makes complete isolation possible, at least as an asymptotic limit. Let us call systems so behaved "C-systems". Then, there are objects which behave as quantum mechanics say they should ("quantons" in Bunge's terminology). These are very different: a quanton never presents sharp ascertainable values for all its applicable dynamical properties, its possible states include combinations of different values for any given applicable property, full isolation is not possible for quantons that have interacted in the past, and a quanton's description is irreducibly probabilistic.

So, the world seems to present us with at least two very different kinds of objects. There is nothing wrong in principle with this. But, as Schroedinger famously realized, if microscopic objects such as atoms can be in strange superpositions, so can macroscopic objects, because they are made of atoms and superpositions initially confined to microscopic systems can easily creep up into the world of ordinary macroscopic bodies. Unless, of course, there is something special about the encounter of the two kinds of systems, something that restores normality by forcing a drastic abortion of quantum superpositions before they become too bizarre. Otherwise microscopic superpositions should routinely manage to get amplified and creep up as macroscopic superpositions in the world of ordinary experience. Nothing of the sort happens, of course, so why does the world behave like this? The standard theory (SQT) bluntly solves this problem by presenting the measurement process as a radical reducer of bizarre quantum mechanical superpositions. According to this view, a measurement process prompts a chance selection of one of the possible values for the magnitude being measured, with the probability of obtaining a particular value given by Born's Rule. But, what counts as a "measurement"? And exactly how does measurement manage to do impose this peculiar form of change? The theory keeps uncomfortably silent about these important matters. At least part of the reason why SQT is so crude about the measurement process is that the part of its dynamics with clearest physical motivation, the purely linear part of the theory, simply cannot resolve superpositions. The linear part is both explicit and resourceful, and one can use it to follow the evolution of a measurement process. However, once a quantum theoretic description of a measurement apparatus is put in place, it is no longer possible to speak of a measurement event occurring, since this would involve reference to a measurement apparatus; neither can one speak of the measurement apparatus as having all its applicable properties informationally available, for the quantum mechanical description only allows one to speak of dispositions. So, are quantons different in kind from ordinary objects, even though the latter are manifestly made up of a great number of interacting quantons? If so, where does the one type of system end and the other begin? Quantum mechanics began as an answer to questions in microphysics, but it turned out to have much to say about questions ranging from the nature of macroscopic systems to even the early state of the universe.

Opponents to granting universality to quantum theory try to limit it to systems with comparatively few degrees of freedom. However, once quantum mechanics is granted free range in microphysics, what considerations prevent one from giving a quantum mechanical account of the emergence of the ordinary, "classical" world? Here many thinkers side with Bohr at his most extreme and simply declare that our understanding is conceptually constrained by classical physics. At least since the 1950s, however, most supporters of the standard theory have been increasingly open-minded about the relationship between the quantum and classical realms. Unlike their most orthodox counterparts, these thinkers take the measurement processes as something derivative rather than fundamental, and they conceive of the standard postulate about wave function collapse as just a "first approximation" model. Those who, in addition, consider wave function collapse an objective aspect of nature, take measurement contexts as situations that merely take full advantage of naturally occurring collapse events. It was in this vein that many objectivist models of the interaction between quantons and ordinary macroscopic bodies got launched with increasing vigor from the 1950s on[1].

All along, however, the unification project has proceeded against a background of expectations in the form of a list of theses that, it seems, any prospective conceptual unification with classical physics must be able to establish if it is to be minimally respectable.

---

[1] The projects in question prominently include Bohm 1952 and Everett 1957.

2. A widely shared list

The desiderata that follow, articulated by R.I.G. Hughes, expresses the consensus of a wide and influential group of foundationists at the dawn of the 1990s about the claims that ought to be derivable from an adequate unification theory[2]. The claims in question centrally include the following:

1. A C-system behaves like a large composite q-system.

2. Differences between quantons and C-systems are attributable to the complexity of the latter.

3. Classical behavior emerges with the massive increase of complexity in a quantum system.

4. Some systems are large enough to be regarded as C-systems.

5. In an interaction between a C-system and a quanton, some properties of the former are realized probabilistically in accordance with Born's Rule.

6. In such an interaction, together with the realization of a particular property of the C-system, there comes a corresponding localization of the state of the quanton.

Hughes ventures no explanation of the differences between C-systems and quantons; to him, as with many other foundationists, theses 1-6 constitute a list of the problems an account of the classical-quanton relation would have to resolve. Hughes also shares some Copenhaguen-like concern about the conceptual integrity of current quantum theory. In particular, he thinks that the explanation of how we can use the limiting cases of quantum theory in order to formulate the theory cannot be given within the theory itself–that explanation, he believes, will have to wait the arrival of a new physical theory, one which is not formulated against a "classical horizon" in the way that quantum theory is.

Be that as it may, there is little question that the above desiderata has helped the search for better formulations and interpretations of quantum theory, especially through such developments as the analysis of

---

[2] Hughes 1989, pp 316-17.

measurement processes in terms of environment-induced decoherence, more coherent formulations of Bohmian mechanics, and strictly physicalist modelling of wave function collapse. In order to make the discussion of the desiderata's import clearer I will focus on just one of these particular approaches, the GRW program. My interest here is not in evaluating this program against other contemporary contenders, let alone endorsing it especially[3]; I simply want to assess how one major approach responds to the challenge posed by the desiderata, and then see how its manifest shortcomings in this regard contribute to the charge that the unification of classical and quantum mechanical systems at hand is of merely instrumental value. This is of interest because the other mentioned programs seem liable to very similar charges.

3. A representative example of response to the challenge

The GRW approach—introduced by Ghirardi, Rimini and Weber in the mid-1980s, and subsequently developed by many others[4]—postulates a fundamental mechanism of spontaneous state reduction that operates in the position basis. For a single, isolated microscopic particle, the stochastic process assumed by the theory is extremely weak: every 10^{16} seconds (10^9 years) or so, on average, the particle's state is overtaken by a sudden contraction in the coordinate representation. Systems with more degrees of freedom develop kicks proportionately. For a system of N distinguishable particles, localization of the i-th particle is effected by the operator:

$$L_x{}^i = (/)^{3/4} \exp[(-q^i - x)^2/2], \qquad (1)$$

where $L_x{}^i$ is a norm-reducing, positive, self-adjoint operator in the corresponding N-particle Hilbert space, representing the localization of the i-th particle around the point x. Kicks affecting the i-th particle occur with a mean frequency $_0$, each particle acting as an independent source of kicks, the frequency of kicks for the total system thus amounting to $_{i0} = N_0$. The

---

[3] I present a critical review of the GRW approach in Cordero 1999.

[4] See especially Ghirardi, Rimini and Weber 1986, Ghirardi and Pearle 1992, Ghirardi, Grassi, Butterfield and Fleming 1992. For a mass-density formulation of the approach, see Ghirardi, Grassi and Benatti 1995.

postulated probability density for a localization of the i-th particle at a location x, given a total wave function, is

$$Pr_i(x) = L_x^i 2, \qquad (2)$$

which requires $d^3x \ (L_x^i)^2 = 1$. Between successive spontaneous collapses the dynamical evolution of the total state is exactly as QT_0 prescribes.

From a GRW perspective, what was previously an odd and inexplicable difference of behavior between standard microscopic and macroscopic systems now follows naturally from the dynamics. An ordinary laboratory pointer, for example, contains about 10^{20} atoms, and so a kick is expected to develop in it every 10^{-4} seconds, on average. Indeed, whenever the center of mass of an approximately rigid N-particle system enters a superposition involving a position spread greater than 1/, that superposition is aborted in $10^{16}/N$ seconds, on average. But, does this proposed resolution of the measurement problem fulfill the desiderata profiled in the previous section?

4. Grw and the list

The GRW theory purports to model a standard macroscopic body (a C-system) as one that behaves like a large composite system of quantons, as thesis 1 requires. It achieves this by planting the seeds of classical determinateness at the most universal level of the dynamics. All the other desiderata are handled with similar ease. The GRW approach traces the differences between systems of quantons and C-systems directly to the complexity of the latter, as thesis 2 requires. And the theory presents C-behavior as emerging with an increase of complexity, as thesis 3 requires; the proposal further shows how some systems are large and complex enough to be regarded as C-systems (thesis 4). The GRW model seems also clear about how, in an interaction between a C-system and a quanton, relevant properties of the former are realized probabilistically (thesis 5). Thesis 6 is also honored: when such an interaction occurs, together with the realization of a particular property of the C-system, there comes a localization of the state of the quanton counterpart. All this may be, perhaps, best appreciated in concrete terms by seeing how the GRW theory handles the evolution of bulky systems, say a typical laboratory pointer.

A high level of preservation of the internal structure of ordinary macroscopic bodies by the collapse process is crucial for the approach to work (it would be no good if the kicks mangled ordinary objects). This, however, does not seem to be a problem for the model at hand. Given a set $\{A_m>\}$ of, say, pointer states such that the distance between any two intended pointer positions A_m is much greater than $1/$, the state of a definite pointer position is as follows:

$$A_m> = \text{CM wave function}> \text{ internal structure wave function}>$$
$$= (X_{A(m)})>_{CM}(q_i)>_{INT} \qquad (3)$$

For a rigid structure like a pointer, the function is very sharply peaked in its (internal) coordinate representation—i.e., boundaries are always clearly defined with respect to coarse-graining of order $1/$. The crucial question is what happens to the internal structure (represented by) during a GRW collapse. Happily for the approach, its proposed nonlinear and stochastic part of the dynamics preserves internal structure. Specifically[5]:

$$L_X^A \{(X_{A(m)})>_{CM}(q_i)>_{INT}\} = \{L_X^A (X_{A(m)})>_{CM}\}(q_i)>_{INT}$$

$$q_i = Q + q'_i(r). \qquad (4)$$

By the approximate rigidity condition, the structural wave function can be assumed to be peaked around the internal structure vector r_0, so that:

$$L_X^i[(Q)>(r)>] \ (/)^{3/4} \exp\{-[Q + q'_i(r_0) -x]^2/2\} (Q)>(r)>]$$

$$= (/)^{3/4} \exp[-[Q -C_i)^2/2]$$

$$= [L_{C-i}^{CM}(Q)>](r)> . \qquad (5)$$

This establishes two significant results. Firstly, a GRW localization process does not affect the internal structure of an approximately rigid pointer. Secondly, localization of a single component particle in the latter forces into localization the center of mass. Since, by hypothesis, particles remain subject to localization events independently of their state of agreggation, $^{cm} = {}^1_i$.

---

[5] See, in particular, Ghirardi and Rimini 1990.

More than structural preservation is needed, however. The description of a C-system calls for a register of actual properties that no "pure" quantum mechanical state can provide. The GRW model deals with this difficulty by showing the power of the natural amplification effects available to its spontaneous collapse mechanism. Sufficiently complex GRW systems "always have a definite position", because the collapse mechanism keeps their position sharp within the margin characteristic of GRW wave function reductions (quantified by the parameter). And, to the extent that their position is sufficiently continuous, velocity and momentum get to be similarly well defined, along with all the classical dynamical properties defined in terms of them. The resulting classical state is not "zero-grain", to be sure, but just extremely sharp relative to ordinary levels of experimental error, with no bizarre spread allowed to last more than "instant" with any significant probability. This leads to constraints on quantum mechanical spreads so stringent for any C-system that all the corresponding quantum averages become effectively punctual, which transforms the relations provided by Ehrenfest's theorem into a coarse-grained version of the relevant classical laws.

So, in the GRW model, C-systems enjoy a significant level of self-sustaining classicality. But what happens when a C-system interacts with a simple quanton? When a system complex enough in terms of particle numbers to be regarded as a C-system—let us call it "M"—encounters a quanton, the entanglement dictated by the linear part of SQT quickly leads to the rise of Schroedinger-cat-like states involving M. However, these can only be expected to last exceedingly little, because of the material constitution of M, which virtually guarantee the development of a GRW "kick" on one of its constitutive particles well before any bizarre phenomenon creeps into the ordinary world. The embrionic Schroedinger-like-cat situation is overcome by a massive raise of the relative amplitude of one of the original terms, which effectively reduces the total state along the lines of the old collapse postulate (now grounded in a law that picks a preferred basis in nature).

And so, the GRW approach does seem to go a long way toward satisfying the six major desiderata left by SQT. Why, then, isn't everybody happy?

5. Not classical enough?

All the above results nothwistanding, it seems clear that in the GRW approach C-systems cannot free themselves from many seriously non-classical features. One such feature is the persistence in the quantum state of classically incompatible terms after the selection process. This occurs because the amplitude of an initially competing term can never go to zero in a theory that preserves the linear part of SQT. The potential distress here is compounded by the way in which the specific GRW collapse mechanism merely compresses the wave function around one of the initial terms, without altering the part that deals with internal structure, as the pointer example shows. To many thinkers this yields unpalatable consequences in Schroedinger-cat-like situations. In particular, the separability secured for complex systems by the GRW model is "effective" rather than strict, and so Schroedinger-like-cat situations are not really annihilated but merely rendered inscrutable "for all practical purposes", given our coarse-grained empirical access to physical systems in general. This makes GRW property states effectively rather than strictly classical, and then only up to the level of coarseness imposed by the parameter of the theory. Consider the transition prompted by the GRW mechanism amounts essentially to the following:

GRW Schroedinger-cat-like entanglement>
Term favored by selection> + All the other terms>,   (6)

where stands for a value extremely close to zero. One problem with the above evolution is that, precisely because the GRW mechanism changes only the center of mass part of the wave function, the term with negligible amplitude (the GRW "tail") keeps all the "internal life and vibrancy" it had before the selection occurred. That is, the array of alternative histories it represents keep unfolding as if nothing had happened.

Defenders of GRW tend to overlook this feature of the proposal, but the suggestion here is that the dynamics keep wave function tails "too active". In particular, if the GRW wave function is interpreted realistically, then it becomes extremely difficult to understand why the histories that develop in the tails should not be considered as *fully real* as either the wave function itself or the history associated with its dominant part in coordinate representation. One charge against the GRW approach is, therefore, that its picture of the C-world looks too much like a special case of the "Many

Worlds" picture—with a dominant branch[6]. Happily, this problem about the lively character of the tails seems an artifact of the specific mechanism favored by GRW: not all imaginable collapse mechanisms present a similar inconvenience[7]. However, all collapse models must allow for wave function tails, as this is a feature imposed on quantum mechanical models that preserve the linear part of the SQT dynamics.

I have rehearsed two different kinds of issues regarding the suitability of the GRW proposed unification of classical and quantum models. The first hinges on specific difficulties faced by the GRW model for ordinary macroscopic objects, namely that the approach seems to end up too close to its supposed nemesis, the many worlds theory. The second issue is more uncompromising: in either SQT or the GRW approach the term "position" and most other relevant terms differ in meaning from their homonymous classical counterparts. To partisans of this line the highlighted coarseness of the GRW property state is only the tip of the iceberg, especially if quantum mechanical observables are defined in Hilbert space terms, and so ultimately in terms of vectors and their superpositions. (Classical properties, by contrast, are defined in terms of phase space and do not admit of superpositions). The point is that no classical system can ever be properly embedded in the conceptual framework of quantum mechanical "C-systems". Moreover, this form of unbrigeability clearly permeates all the best unification projects in contemporary philosophy of physics. It certainly affects the decoherence approach, and, except for the position of particles, also Bohmian mechanics.

No quantum system can be "truly classical", whatever its number of particles or complexity. The question is: how much does this matter?

6. The classical world in qt

One influential reaction to the above difficulties is to partially revert to empiricist lore and accept that quantum theory occupies a very unusual place among physical theories—particularly to the view that quantum theory contains classical mechanics as a limiting case, yet at the same time it *requires* this limiting case for its own formulation. This doctrine, rooted in

---

[6] Cordero 1999.

[7] Cordero 1995.

the Copenhagen school, has been given philosophical stamina in recent times by such important philosophers as R.I.G. Hughes and many others, especially in the formalist camp[8]. But, exactly how is classical physics required by quantum physics?

Two methodological verities come immediately to mind. One is that quantum mechanical predictions require the use of devices that satisfy classical physics to a high degree of approximation. Thus, for example, spectroscopic data can only be gathered with the help of apparatus whose behavior is eminently classical. No serious case for "primacy" seems apparent here, however, for the classical behavior involved can be described as C-classical just as adequately.

The second verity has to do with the way in which quantum physics is heuristically linked to classical physics. As a matter of historical fact, most of the non-relativistic Hamiltonians in quantum theory have been modelled with the help of crucial structural inspiration from classical physics. But, again, this level of primacy seems as philosophically uninteresting as the previous one. All sensible reasoning—not just in physics—approaches its intended domain using as heuristic background prior information about it.

There is, however, a seemingly deeper way of trying to grant primacy to at least one major aspect of the classical world, a way still very popular in formalist quarters. It amounts to the claim that physical representation in general should respect the classical (Boolean) structure of the set of events associated with any given observable. Why? Because, the story goes, all special fields of inquiry must be accessible from some common general background of human life. To many contemporary thinkers not only is that background centered around ordinary observation but observation is essentially linked to the Boolean lattices characteristic of classical observables[9].

I think this last position simply begs the question against quantum theory and open physical theorizing in general. As far as anyone can reasonably tell, Boolean lattices provide no more than adequate models for the properties of ordinary macroscopic objects. Beyond this, much argumentation would be needed in order to grant philosophical primacy to Boolean lattices or indeed to any specific structure from classical mechanics. It is not enough just to appeal to the epistemological importance of observation. For, as the critique of strong empiricism has taught us, no level

---

[8] See Hughes 1989, especially Chapter 10.
[9] Hughes 1989.

of observation relevant to science is either unproblematic with respect to truth nor independent with respect to all theory. We should be thus wary of granting sacrosanct status to anything in science, for no level of physical description is closed in principle to the possibility of deep structural transformation as a result of learning.

If so, solving the measurement problem and unifying classical and quantum mechanics into a single theory cannot be conditioned to the exact preservation of any particular lore from either classical theory or first philosophy. No exact embedding of the models of one theory into those of another may be possible, but none seems actually required for descriptive advance—unless, again, one begs the question in favor of some unwarranted first philosophy. As Roberto Torretti so lucidly noted some time ago in connection with the issue of theory change, "there is an inevitable fuzziness in the way each domain is inserted or 'embedded' in the background—this, in turn, favors the gross identification of some objects referred to by diverse theories, even if the latter conceive them very differently" (Torretti 1990, p. 79).

References

Bohm, D.: 1952, "A Suggested Interpretation of the Quantum Theory in Terms of 'Hiddden Variables'", Parts I and II, *Physical Review* 85, pp. 166-93.

Cordero, A.: 1995, "A GRW-Like Approach to the Measurement Problem", in M.L. Dalla Chiara (ed.): 1995, *10th International Congress of Logic, Methodology and Philosophy of Science*. Florence: International Union of History & Philosophy of Science, pp. 454-55.

Everett, H.: 1957, "Relative State Formulation of Quantum Mechanics", *Reviews of Modern Physics* 29, III, pp. 454-62.

Ghirardi, G., Grassi, R., and Benatti, F.: 1995, "Describing the Macroscopic World: Closing the Circle within the Dynamical Reduction Program", in *Foundations of Physics* 25, pp. 5-38.

Ghirardi, G., Grassi, R., Butterfield, J. and Fleming, G.: 1992, "Parameter Dependence and Outcome Dependence in Dynamical Models for Statevector Reduction", *Foundations of Physics* 23, pp. 341-364.

Ghirardi, G. and Pearle, P.: 1992, "Elements of Physical Reality, Nonlocality and Stochasticity in Relativistic Dynamical Reduction Models", in A. Fine, M. Forbes, and L. Wessels (eds.), *Philosophy of Science Association, Vol. Two*. East Lansing, MI: Philosophy of Science Association, pp. 35-48.

Ghirardi, G., Rimini, A. and Weber, T.: 1986, "Unified Dynamics for Microscopic and Macroscopic Systems", *Physical Review D* 34, pp. 440-491.

Ghirardi, G. and Rimini, A.: 1990, "Old and New Ideas in the Theory of Quantum Measurement", in A.I. Miller (ed), *Sixty-Two Years of Uncertainty*. New York: Plenum Press, pp. 167-191.
Hughes, R.I.G.: 1989, *The Structure and Interpretation of Quantum Mechanics*. Cambridge, MA: Harvard University Press.
Torretti, R.: 1990, *Creative Understanding: Philosophical reflections on Physics*. Chicago: University of Chicago Press.

9. Reduction, Integration, Emergence and Complexity in Biological Networks

Jacques Ricard

Institut Jacques Monod, Universités Paris 6, Paris 7, CNRS

Modern science, that is, the scientific activities that have sprung up since the time of Descartes, Newton and Leibnitz, has always be in search of simplicity. It is based upon the four "principles" expressed by Descartes[1] in the *Discours de la Méthode*. *"Le premier était de ne recevoir jamais aucune chose pour vraie, que je ne la connusse évidemment être telle: c'est-à-dire d'éviter soigneusement la précipitation et la prévention; et de ne comprendre rien de plus en mes jugements, que ce qui se présenterait si clairement et si distinctement à mon esprit que je n'eusse aucune occasion de le mettre en doute.*

Le second, de diviser chacune des difficultés que j'examinerais, en autant de parcelles qu'il se pourrait, et qu'il serait requis pour les mieux résoudre.

Le troisième, de conduire par ordre mes pensées, en commençant par les objets les plus simples et les plus aisés à connaître, pour monter peu à peu, comme par degrés, jusques à la connaissance des plus composés; et supposant même de l'ordre entre ceux qui ne se précèdent point naturellement les uns les autres.

Et le dernier, de faire partout des dénombrements si entiers, et des revues si générales, que je fusse assuré de ne rien omettre".

The second principle is no doubt reductionist in its essence and, for this reason, antagonistic to present studies of complex systems. In spite of the fact that these analytic and reductionist approaches have led to important discoveries, there is little doubt that the world is complex in its essence and

---

cannot be correctly understood through the modern reductionist approach. This is precisely what has recently become evident in different fields of science ranging from fundamental physics to the social sciences[2]. It is clear that the living creatures on earth are indeed complex, and it is hopeless to try to understand "what life is" solely through the study of the macromolecules that constitute the substance of these living creatures.

The aim of the present contribution is threefold:
- first, to present a tentative definition of the concepts of reduction, integration, emergence and complexity;
- second, to give a brief overview of the main features of complex systems;
- third, to discuss more thoroughly the most important of these features, the one which is at the very basis of complexity and can be used to define this concept, namely information.

Although this discussion will be general, it will be more specifically oriented toward dynamic biological systems.

1. A tentative definition of reduction, integration, emergence and complexity

From a philosophical viewpoint, the term "reduction" can have at least two different meanings. It may refer to the mental process of deriving one scientific theory from another, more general and embracing, theory. In this perspective, a biological theory for instance could be reduced to a more general physical theory. If such a reduction could be pursued *ad infinitum* this would imply the unity of science.

But there is a second type of reduction that is directly related to the problem of emergence and complexity. Let us consider a system consisting, as all systems do, of a number of sub-systems. One can expect three types of situations. First, the overall system and the component sub-systems have the same degrees of freedom, or the same entropy, or the same properties. In this case, the overall system is not a real one, for it simply corresponds to the juxtaposition of different sub-systems. It can therefore be reduced to its

---

[2] See Gallagher and Appenzeller 1999, Goldenfeld and Kadanoff 1999, Whitesides and Ismagilov 1999, Hatwell et al. 1999, Sethna et al. 2001, Ricard 1999.

components. The second possible type of situation occurs if the overall system has fewer degrees of freedom, or a lower entropy, than the set of component sub-systems. It thus behaves as a real system, for it displays some sort of integration of its components in the form of a coherent whole. The integrated system therefore evinces collective properties distinct from those of the component sub-systems. The last possibility is observed if the overall system has more degrees of freedom, or more entropy, than the set of component sub-systems. One may expect this system to be richer, and to have more collective properties than the simple integrated system considered above. It can be defined as an emergent, or complex, system.

As the concepts of integration and emergence are related to the concept of information, it is thus clear that the analysis of this concept should be central in any discussion of complexity.

2. The main features of complex systems

Before discussing at length the logical foundations of the concepts of reduction, integration and emergence, it is of interest to present briefly the main features of complex systems.

- A complex system should possess information. This matter will be discussed later.
- A complex system exhibits a certain degree of order. It is neither strictly ordered nor completely disordered.
- A complex system should display collective properties, that is properties different from those of the component sub-systems.
- A complex system is not in thermodynamic equilibrium, it is thermodynamically open and displays nonlinear effects.
- A complex system is in a dynamic state and has a history. This means that the present behavior of the system is in part determined by its past behavior.
- A complex system has emergent collective properties.

3. Information, integration and emergence in biological complex networks

The total information of the living cell is often identified with its genetic information. This is too restrictive a view of the concept of biological information. As a matter of fact, most biological networks can probably act as information channels, or as generators of information. Classical Shannon information theory, however, is not ideally suited to describing what network information is. It has therefore to be revisited and altered in order to allow this study. Moreover many papers have recently been devoted to networks, in particular to metabolic networks. Before studying network information, integration and emergence, one has to present a brief overview of what a biological network is.

3.1. Biological networks

A network is a set of nodes connected in accordance with a certain topology. In a metabolic network, for instance, the nodes are the various metabolites, or the enzyme-metabolite complexes, and the edges connecting the nodes the corresponding chemical reaction steps. Networks belong to different types known as random, regular, small-world and scale-free[3]. In random graphs, the nodes are connected according to a Poisson distribution, whereas in regular networks the node connection is effected through a fixed topological rule. Small-world networks display a fuzzy topology, midway between pure randomness and strict regularity. Scale-free graphs have both poorly and highly connected nodes.

A network is roughly described by a parameter called diameter. If we consider all the possible pairs of nodes in a graph, and the shortest possible distance between these pairs, this distance is the smallest number of steps that separate the two nodes of all the possible pairs; and the network diameter is the mean of all the shortest distances. The immediate consequence of this definition is an increase of the diameter of a random graph as the number of nodes increases.

Metabolic networks possess an interesting property that is worth discussing briefly[4]. Sequencing genomes of very different living systems,

---

[3] See Strogatz 2001, Watts and Strogatz 1998, Albert et al. 2000, Jeong et al. 2000, Fell and Wagner 2000.

[4] See Jeong et al. 2000.

ranging from mycoplasms and bacteria to man, has allowed us to know all the enzymes present in these organisms and therefore all the reactions they catalyze. As a consequence, this allows us to construct all the metabolic networks of these organisms. Depending whether the living organisms are "simple" or "complex", the numbers of nodes of these graphs are very different. Still the graph diameter remains constant along the phylogenetic tree. The reason for this constancy is that metabolic networks are not random, but possess a fuzzy organization. They are of the small-world type, and are more connected the more "complex" the organism.

3.2. The sub-additivity principle, reduction and integration

Let us consider a composite dynamic network made up of four classes of nodes. A first class, X, collects the nodes associated with a discrete variable x. A second class, Y, describes a different property associated with a different discontinuous variable y. A third class, XY, has both properties x and y, and a final class has none of them.

The nodes of X and Y have a certain probability of occurrence $p(x)$ and $p(y)$, respectively. Similarly, the nodes of class XY have probabilities $p(x,y)$. If the events x and y are independent, then

$$p(x,y) = p(x)p(y) \qquad (1)$$

But if the events x and y are correlated, then

$$p(x,y) = p(x)p(y/x) = p(y)p(x/y) \qquad (2)$$

which is the classical Bayes' relationship. Here, $p(x/y)$ and $p(y/x)$ are conditional probabilities, that is the probability of a value of x given a value of y and the probability of a value of y given a value of x, respectively. At least for correlated events, conditional probabilities are usually larger than the corresponding probabilities. This is important, for it represents the very basis of what is termed the sub-additivity principle[5], which will be discussed later on.

The concept of information is related to that of uncertainty. The larger the uncertainty of a message, the larger its information content. Put in other words, the information of a system is its ability to perform a difficult, and

---

[5] See Shannon 1948, 1949, Yokey 1992, Adami 1998.

therefore improbable, task. The nature of information should thus be related to that of uncertainty. Any mathematical function aimed at measuring the degree of uncertainty should meet the requirements of two axioms: that of monotonicity and that of additivity. Monotonicity means that the uncertainty should increase regularly with the number of states of the variables x or y. Additivity should express the view that if the discrete variables x and y are independent, the uncertainty of the pairs XY should be equal to the sum of the uncertainties of X and Y. The only simple function of a probability p that meets these two axioms is

$$f = \log(1/p) \qquad (3)$$

The corresponding mean uncertainty function H of probabilities p is therefore

$$H = -\langle \log p \rangle \qquad (4)$$

where log is taken to mean the logarithms to base 2. One can thus define mean uncertainty functions (or entropies expressed in bits) of X and Y as

$$H(X) = -\langle \log p(x) \rangle$$
$$\qquad (5)$$
$$H(Y) = -\langle \log p(y) \rangle$$

Similarly, one can also derive the expression of the mean uncertainty function of the pairs XY (or joint entropy), namely

$$H(X,Y) = -\langle \log p(x,y) \rangle \qquad (6)$$

Simple inspection of equations (5) and (6) shows that if the events X and Y are independent then

$$H(X,Y) = H(X) + H(Y) \qquad (7)$$

This equation shows that the entropies, or the degrees of freedom, of the system XY and of the sum of the sub-systems X and Y are the same. Equation (7) therefore implies that the system XY can be *reduced* to its component sub-systems X and Y.

Now, if the events and the corresponding variables x and y are correlated, then one can demonstrate that the mean value of conditional probabilities is larger than the mean of the corresponding probabilities. This implies that

$$<\log p(x/y)> > <\log p(x)> \qquad (8)$$

and

$$<\log p(y/x)> > <\log p(y)> \qquad (9)$$

and this leads to

$$H(X,Y) < H(X) + H(Y) \qquad (10)$$

Therefore the sub-additivity condition can be expressed as

$$H(X,Y) \leq H(X) + H(Y) \qquad (11)$$

and the difference, $I(X:Y)$, between the two members of this expression is the information of the network. Therefore

$$H(X,Y) + I(X:Y) = H(X) = H(Y) \qquad (12)$$

Information is therefore the part of entropy that should be added to the joint entropy $H(X,Y)$ in order to obtain the sum of individual entropies $H(X)$ + $H(Y)$. In the case of correlated discrete variables X and Y, this information is of necessity positive and expresses the degree of integration of system XY with respect to its component sub-systems X and Y[6]. Put in other words, expression (10) shows that an integrated system has less entropy, or fewer degrees of freedom, than the two component sub-systems X and Y.

3.3. Lack of sub-additivity and emergence

The above reasoning and the sub-additivity principle, which is at the heart of Shannon's information theory, are valid only because the discontinuous

---

[6] See Tononi et al.1994.

variables are correlated. If, however, these variables represent events that physically interact, then the sub-additivity principle does not necessarily hold. This situation can be illustrated in simple enzyme-catalyzed reactions. Let us consider the group transfer chemical reaction

$$AX \; + \; B \longrightarrow \; A \; + \; XB.$$

If this reaction is catalyzed by the enzyme E, this means that AX and B can form binary complexes with the enzyme, namely

$$E \; + \; AX \; \underset{}{\overset{K_A}{\rightleftharpoons}} \; E.AX \quad \text{and} \quad E \; + \; B \; \underset{}{\overset{K_B}{\rightleftharpoons}} \; E.B$$

Once these binary complexes are formed, another molecule of B and AX can be added to E.AX and E.B, respectively. This leads to the appearance of the ternary complex E.AX.B in the reaction mixture. One has

$$E.AX \; + \; B \; \underset{}{\overset{K'_B}{\rightleftharpoons}} \; E.AX.B \quad \text{and} \quad EB \; + \; AX \; \underset{}{\overset{K'_A}{\rightleftharpoons}} \; E.AX.B$$

Catalysis takes place on the ternary complex which decomposes, leading to reaction products, namely

$$E.AX.B \; \overset{k}{\longrightarrow} \; E \; + \; A \; + \; XB$$

Let us assume for the moment that the overall system is close to thermodynamic equilibrium, which implies that the rate constant of the decomposition of the ternary complex is small. Then one can derive the expression of the probabilities for the enzyme to bind AX, p(AX), or to bind B, p(B), or to bind both AX and B, p(AX,B). Similarly, one can derive the expressions of conditional probabilities p(AX/B) and p(B/AX). One notices that the probabilities p(AX) and p(B) not correlated to the concentration of AX and B have a fixed value in the reaction medium, which implies that the probabilities p(AX) and p(B) assume only one value. But one can still extend the concepts of entropy and information to the present situation by setting

$$H(AX) = - \log p(AX)$$
$$H(B) = - \log p(B) \qquad (13)$$
$$H(AX,B) = -\log p(AX,B)$$

These expressions are indeed identical to equations (5) and (6) if X, Y and XY each assume only one value. Expressions of these probabilities (equation 13) are dependent upon the concentrations of the reagents AX and B, as well as upon the affinity constants K_A, K'_A, K_B and K'_B. One can demonstrate that if $K_A = K'_A$, which implies from thermodynamics that $K_B = K'_B$, then

$$p(AX/B) = p(AX)$$

$$p(B/AX) = p(B) \qquad (14)$$

Under these conditions

$$H(AX,B) = H(AX) + H(B) \qquad (15)$$

and the enzyme network does not contain any information. The corresponding system of joint entropy H(AX,B) can be reduced to its sub-systems of entropies H(AX) and H(B).

Alternatively, if $K'_A > K_A$ and $K'_B > K_B$, then

$$p(AX/B) > p(AX)$$

$$p(B/AX) > p(B) \qquad (16)$$

and

$$H(AX,B) < H(AX) + H(B) \qquad (17)$$

The system behaves as an integrated coherent entity. Its entropy, H(AX,B), is smaller than the sum of the entropies of the two sub-systems H(AX) and H(B). The information of the network, I(AX:B), is thus positive and equal to the extent of the integration process. The system behaves as an information channel.

The last case is the most interesting. If $K_A > K_A'$ and $K_B > K_B'$, then

$$p(AX) > p(AX/B)$$

$$p(B) > p(B/AX) \qquad (18)$$

and

$$H(AX,B) > H(AX) + H(B) \qquad (19)$$

The system behaves as an emergent coherent entity. Its entropy is now larger than the sum of the entropies, $H(AX)$ and $H(B)$, of the corresponding sub-systems. It can therefore be considered a complex system. The enzyme network does not act as an information channel but generates information of its own. Here information, $I(AX:B)$, defined as a measure of the extent of the emergent process, has a negative value. It is the piece of entropy that should be deduced from the joint entropy $H(AX,B)$ in order to obtain the sum of the individual entropies $H(AX)$ and $H(B)$.

As the sub-additivity principle lies at the root of standard information theory [13-16], the last conclusion which has been discussed above is at variance with this theory. It is therefore important to know what is basically different in the present reasoning with respect to the one followed by Shannon. Classical Shannon theory is based on the concept of correlation between discrete variables without reference to any molecular physical interaction. The sub-additivity principle should of necessity apply under these conditions. In the situation discussed above, which is most likely the one taking place for most biological networks, there is no correlation but a physical interaction between two binding processes. Under these conditions, sub-additivity may, or may not, apply. If $K_A' > K_A$ and $K_B' > K_B$, the binding of B to the enzyme facilitates that of AX, and conversely. This situation should conform to sub-additivity. But if $K_A > K_A'$ and $K_B > K_B'$, then the binding of B hinders that of AX, and conversely. In this case, the sub-additivity principle no longer applies.

So far, the simple enzyme network was assumed to exist under quasi-equilibrium conditions. If, however, the system departs from this state, that is if the rate constant k assumes rather high values, the system can reach a new steady state. Under these conditions, the probabilities for the enzyme to bind AX, B or both AX and B depend on the value of k. As a matter of fact, the probabilities p(AX), p(B) and p(AX,B) decrease as k is increased, but p(AX,B) falls off much more steeply than p(AX) and p(B) in such a way that H(AX,B) becomes larger than H(AX) + H(B). Moreover this situation takes place even if $K_A = K_A'$ and $K_B = K_B'$, that is, even if the binding of the two substrates do not physically interact. This means that a departure from quasi-equilibrium results in the emergence of information.

4. General conclusions

The aim of the present contribution has been to offer a clear-cut and mathematically sound definition of the concepts of reduction, integration and emergence through the use of information theory. As these effects are likely to take place in most biological networks, whatever their nature, Shannon's information theory has been revisited and modified so as to apply to these networks.

If two, or several, discrete variables referring to the properties of a quasi-equilibrium network are correlated, then the sub-additivity principle should apply and the system can adopt two different types of behavior. If its joint entropy is equal to the sum of individual entropies of component sub-systems, the network properties can be reduced to those of the sub-systems. If, alternatively, the joint entropy of the overall network is smaller than the sum of individual entropies of the component sub-systems, the network behaves as an integrated entity whose properties cannot be reduced to those of its elements. The information content of the system is a measure of its degree of integration.

If, in a quasi-equilibrium network, two or several events are not correlated but physically interact, the system may undergo either reduction or integration, as described above. But it may also display emergent properties if its joint entropy is larger than the sum of the individual entropies of its component sub-systems. Under these conditions, the system does not act as an information channel but generates information of its own. The extent of this information is a measure of the extent of emergence. The origin of emergence is, in the present case, to be found in the physical interactions that take place between two or several molecular events.

If the network does not exist under quasi-equilibrium but under steady state conditions, the departure from equilibrium in itself generates the emergence of information. This situation may occur even in the absence of physical interactions between molecular events. It has its origin in the fact that the system receives and transfers molecular signals coming from the outside. It therefore appears that emergence in networks can occur as a consequence of physical interactions between molecular events and (or) is also induced by a departure from thermodynamic equilibrium.

References

Adami, C.: 1998, *Introduction to Artificial Life*. Springer Verlag.

Albert, R., Jeong, H. and Barbasi, A. L.: 2000, "Error and attack tolerance of complex networks", *Nature* 406, pp. 378-381.

Descartes, R.: 1992, *Discours de la Méthode*. Paris: Flammarion..

Fell, D. E. and Wagner, A.: 2000: "The small-world of metabolism", *Nature Biotechnology* 18, pp. 1121-1122.

Gallagher, R. and Appenzeller, T.: 1999, *Beyond Reductionism*, Science 284, 79.

Goldenfeld, N. and Kadanoff, L. P.: 1999, "Simple lessons from complexity", *Science* 284, pp. 87-89.

Hatwell, L. H., Hopfield, J. J., Leibler, S. and Murray, A. W.: 1999, "From molecular to modular cell biology", *Nature* 402 (sup.), C47-C52.

Jeong, H., Tombor, B., Albert, R., Oltavi, Z. N. and Barbasi, A. L.: 2000, "The large-scale organization of metabolic networks", *Nature* 407, pp. 651-654.

Ricard, J.: 1999, *Biological Complexity and the Dynamics of Life Processes*. Elsevier.

Sethna, J. P., Dahmen, K. A. and Myers, C. R:. 2001, "Crackling noise", *Nature* 410, pp. 241-250.

Shannon, C. E.: 1948, "A mathematical theory of communication", *Bell System Technical Journal* 27, pp. 379-423, pp. 623-656.

— 1949, "The Mathematical Theory of Communication". University of Illinois Press.

Strogatz, S. H.: 2001, "Exploring complex networks", *Nature* 410, pp. 268-276.

Tononi, G., Sporns, O. and Edelman, G. M: 1994, "A measure of brain complexity: relating functional segregation and integration in the nervous system", *Proc. Nat. Acad. Sci.* USA 91, pp. 5033-5037.

Watts, D. J. and Strogatz, S. H.: 1998, "Collective dynamics of 'small-world' networks", *Nature* 393, pp. 440-442.

Whitesides, G. M. and Ismagilov, R. F.: 1999, "Complexity in chemistry", *Science* 284, pp. 89-92.

Yokey, H. P.: 1992, *Information Theory and Molecular Biology*. Cambridge University Press.

Part III

The Emergence of the Mind

10. Complexity and the Emergence of Meaning: Toward a Semiophysics

F. Tito Arecchi

University of Florence and Istituto Nazionale di Ottica Applicata

1. Introduction

In a debate on "Complexity and Emergence", we should first of all provide clear definitions of these terms, in order to ascertain how much of what we say depends on our cultural bias or is an artifact of our linguistic tools, and how much corresponds to hard facts, to our embedding in an open environment the features of which, even though actively elaborated by our semantic memory, can not be taken as sheer "autopoiesis", but are grounded on an ontology.

This inquiry is being made from the point of view of a physicist who has been active for decades in investigating the formation of collective or coherent processes from a large amount of otherwise separate individuals, pointing out the critical appearance (*emergence*) of new world configurations and the elements of *novelty* of this emergence, which make this phenomenon *complex*. By complex we do not mean the trivial fact that the computational cost of their description is high (in such a case I would rather call them *complicated*), but the fact that available knowledge stored in well established models is not sufficient to reliably predict the emergence, and one must integrate the deductive chains with extra information which introduces an historical flavor into the scientific procedure.

This presentation is organized as follows.

In the first part we discuss the sources of wonder, what Plato called the origin of science, that is, why among many peculiarities (*saliences*) we prefer to focus our attention on some (*pregnancies*) (Sec. 2). Then we explore how, as we organize our knowledge in a scientific language, we select the relevant words (*names*) depending on their relation to an ontology (*things*) (Sec. 3).

In the second part we try to put order into the debated issue of *complexity*, introducing a fundamental separation between some purely mental situations without any realistic counterpart (*closed systems*) and what we in fact come across in everyday life (*open systems*) (Secs. 4 and 5).

The third part goes to the very basis of perceptual processes. If we accept—as proper to complex systems—organizing our knowledge over different and mutually irreducible hierarchical levels, each one with its own rules and language, then the most fundamental such level in cognitive processes is the physical description of how external stimuli (light, sound, pressure, chemicals) are transformed into sensorial perceptions.

Already at this neurodynamical level, we come across a quantum limitation which forbids the brain's operations from being fully simulated by a universal computing machine (Sec. 6). I purposely said "brain", since I do not wish to enter the debates about "mind", "consciousness", etc.

I have called "neurophysics" the combination of neurodynamical facts, whereby neurons are treated as physical objects to be compared with lasers or other nonlinear dynamic systems, and the quantum limitation emerging from the peculiar spike synchronization strategy selected in the course of natural evolution as the optimal strategy to elaborate information into relevant cognitive processes.

As for the list of references, I have often replaced the specific mention of an article or a book by a website, where one can conveniently browse for a more satisfactory answer. I think that time is ripe to consider this reference tool as a standard one.

2. Salience vs. pregnancy

The world around us is full of salient features, that is, sharp gradients which denote the transition from one domain to another. Salience can be captured automatically by a scanning detector equipped to grasp differential features. Saliences have a geometric (space-wise) and dynamic (time-wise) flavor. They correspond to objective features: what Thomas Aquinas called "*dispositio rei*" and more recently A. Reinach (a follower of Husserl) called "*Sachverhalt*"[1].

We might say that saliences uncover an ontology[2]; however, in order to classify a set of features and organize them through mutual relations, we need to assign selection criteria. Such descriptive criteria have guided the construction of sectorial ontologies in many AI (Artificial Intelligence) areas[3]. Hence the problem arises: are there individual objects, or is any world organization rather an arbitrary cut that we operate by picking up some saliences and disregarding others?

Historically, the modern European culture, in line with its Greek-Jewish roots, had chosen the first horn of this dilemma; however, contact with Eastern philosophies, through Schopenhauer and Mach, introduced a "conventionalism" or linguistic relativism, whereby one could build different, uncorrelated, ontologies depending on the points of view from which saliences were selected[4].

The recent emphasis on "regional ontologies", which focuses on particular saliences and thence on particular classes of objects, is a modern technical limitation. A philosopher of science[5] would rather say that selecting a point of view gives rise to a particular science, focusing on certain truths rather than others. Yet there is a hard aspect to saliences, that is, they uncover facts that have their own existence, and are not simply dependent on our cultural artifacts.

In line with Gestalt psychology, René Thom has introduced "pregnancy" to denote a subset of saliences which are relevant for the individual observer[6]. In the case of animals, pregnancy is related to vital needs (search of food, escaping from predators, sexual appeal). Some of these needs may be genetically imprinted, certain others are the result of cultural influences. This latter case is particularly important for human beings. In this regard the contribution of J. Piaget called "Genetic epistemology" is fundamental. As one explores the formation of logical structures in children, one realizes that they derive from *actions* on the objects, not from the objects themselves; in other words, the formation of logical structures is grounded in the

---

[1] See Smith: http://wings.buffalo.edu/philosophy/faculty/smith/

[2] See Poli: www.formalontology.it/polir.htm

[3] See Guarino: www.ladseb.pd.cnr.it/infor/Ontology/ontology.html

[4] See Feyerabend 1975 and Capra 1975.

[5] See Agazzi 1974.

[6] See Thom 1988.

coordination of actions, not necessarily in language. In fact, language is one of the possible semiotic functions; the others, such as gestures, or imitation, or drawing, are forms of expression independent of language, as has been carefully studied in the case of deaf-mutes[7]. In any case, Thom insists, against relativism, on the objective character of the prominent saliences, which he classifies in terms of differential geometry[8].

A very convincing dynamic formulation of the emergence of a new feature, or the disappearance of an old one, as a "control parameter" is changed, is given by Landau's 1937 theory of phase transitions[9]. We present the argument in the updated 1973 formulation called "synergetic" by Haken[10], and initially motivated by a new astonishing phenomenon, the laser threshold, consisting of the onset of a collective coherent emission of light out of billions of atoms, which below that threshold instead contribute individual, unrelated (so called spontaneous) emission acts, as occurs in a conventional light source.

Let me anticipate something I'll discuss in greater detail in Sec. 3. Assume for the time being that we succeeded in describing the world as a finite set of N features, each one characterized by its own measured value $x_i (i = 1 \ to \ N)$, x_i being a real number, which in principle can take any value in the real domain $(-\infty,\infty)$ even though boundary constraints might confine it to a finite segment L_i.

A complete description of a state of facts (a "*dispositio rei*") is given by the N-dimensional vector

$$\underset{\approx}{x} \equiv \left(x_1, x_2, ...x_i, ...x_N \right) \qquad (1)$$

The general evolution of the dynamic system $\underset{\approx}{x}$ is given by a set of N rate equations for all the first time derivatives $\dot{x}_i = dx_i / dt$. We summarize the evolution via the vector equation

---

[7] See Evans 1973.
[8] See Thom 1975.
[9] See Landau-Lifshitz 1980.
[10] See Haken 1983.

$$\dot{\underset{\approx}{x}} = \underset{\approx}{f}(\underset{\approx}{x}, \mu) \qquad (2)$$

where the function $\underset{\approx}{f}$ is an N-dimensional vector function depending upon the instantaneous $\underset{\approx}{x}$ values as well as on a set of external (control) parameters μ.

The solution of Eq. (2) with suitable initial conditions provides a trajectory $\underset{\approx}{x}(t)$ which describes the time evolution of the system. We consider as ontologically relevant those features which are *stable*, that is, which persist in time even in the presence of perturbations. To explore stability, we perturb each variable x_i by a small quantity ξ_i, and test whether each perturbation ξ_i tends to disappear or to grow catastrophically.

However complicated the nonlinear function $\underset{\approx}{f}$ is, the local perturbation of (2) provides for ξ_i simple exponential solutions versus time of the type

$$\xi_i(t) = \xi_i(0)e^{-\lambda_i t} . \qquad (3)$$

The λ_i can be evaluated from the functional shape of Eq. (2). Each perturbation ξ_i shrinks or grows in the course of time depending on whether the corresponding stability exponent λ_i becomes positive or negative. The λ_i are called the "local Liapunov exponents".

Now, as we adjust from outside one of the control parameters μ, there may be a critical value μ_c where one of the λ_i crosses zero (goes from + to -), while all the other $\lambda_j (j \neq i)$ remain positive. We call λ_u the exponent changing sign (u stands for "unstable mode") and λ_s all the others (s stands for stable) (see fig. 1).

Around μ_c, the perturbation $\xi_u(t) \approx e^{-\lambda_u t} \approx e^o$ tends to be long-lived, which means that the variable x_u has rather slow variations with respect to all the others, which we cluster into the subset x_s and which varies rapidly. Hence we can split the dynamics (2) into two subdynamics, one 1-dimensional (u) and the other (N-1)-dimensional (s), that is, we rewrite Eq. (2) as

$$\dot{x}_u = f_u(x_u, x_s, \mu)$$
$$\dot{x}_s = f_s(x_u, x_s, \mu)$$
$$(4)$$

The second one being fast, the time derivative $\dot{x}_s$ rapidly goes to zero, and we can consider the algebraic set of equations $f_s = 0$ as a good physical approximation. The solution yields x_s as a function of the slow variable x_u

$$x_s = g(x_u) \qquad (5)$$

We say that x_s are "slaved" to x_u. Placing (5) into the first part of (4) we have a closed equation for x_u

$$\dot{x}_u = f_u(x_u, g(x_u), \mu). \qquad (6)$$

First of all, a closed equation means a self-consistent description, not dependent upon the preliminary assignment of x_s. This gives an ontological robustness to x_u; its slow dependence means that it represents a long-lasting feature; and its self-consistent evolution law Eq. (6) means that we can forget about x_s and speak of x_u alone.

Furthermore, as μ crosses μ_c, a previous stable value $x_u^{(1)}$ is destabilized. A growing ξ_u means that eventually the linear perturbation is no longer good, and the nonlinear system saturates at a new value $x_u^{(2)}$ (see fig. 2).

Such is the case of the laser going from below to above threshold; such is the case of a thermodynamic equilibrium system changing e.g. from a gas to liquid form, or from a disordered to an ordered state, as the temperature at which it is set (here represented by μ) is changed.

To summarize, we have isolated from the general dynamics (2) some critical points (bifurcations: see the shape of fig. 2) where new salient features emerge. The local description is rather accessible, even though the general nonlinear dynamics f may be rather nasty.

Told in this way, the scientific program seems to converge toward firm answers, as compared to the shaky arguments of philosophers. However it was based on a preliminary assumption, that there was a "natural" way of assigning the x_i. In the next section we explore how to extract the x_i from observations.

3. Names and things

To avoid subjective biases, one should replace definitions in everyday language with quantitative assessments. This is done by isolating something which can be represented in a metrical space as a number, and speaking of a larger or smaller degree of it, of the distance between two values etc., by referring to the corresponding numbers.

In modern science this attitude was consecrated by G. Galilei in his 1610 letter to Marc Welser[11], where he says: "don't try to grasp the 'essence' (i.e. don't try to define the 'nature' of things) but stick to some quantitative affections".

For instance, in the case of apples, rather than arguing on the nature of apples, make comparisons among them, based on quantitative measures of such "affections" as their flavor, color, shape, size, taste. I have listed five qualities for each of which we know how to introduce a meter and hence set a metrical space. In Galilei's time, there was a distinction between primary (measurable) and secondary (subjective) qualities. Nowadays, we know how to objectify and measure secondary quality such as flavor or taste; thus, the old distinction is no longer relevant.

Two different attitudes may be adopted, namely,

Phenomenology: once apples are characterized by a sufficient group of parameters, all apples will be a suitable intersection of the flavor axis, the color axis etc. in a multidimensional space; such a description is *complete*

---

[11] See Galilei 1932.

(all apples will be included) and *unambiguous* (two different apples will not overlap in such a multidimensional space; that is, they differ by at least one of their representative numbers).

Reductionism: split the apple into small pieces, and these again into smaller ones, until you reach a set of elementary components (the biomolecules) out of which, with a wise dosage of the elements of the set, one can reconstruct (synthesize) all kinds of apples. This procedure is lengthier than phenomenology, but it is universal; out of a set of components one can also synthesize oranges, dogs etc. Moreover it looks objective; if we come across intelligent beings from elsewhere, we don't know if our selected affections are relevant for them, but surely they know how to split the apple into components and catch each component's dynamics. When the only known interaction law (2) was Newton's, this approach seemed the ultimate one; thus Newtonianism was considered as the new revolutionary approach upon which to build any world view. An Italian writer of the eighteenth century, F. Algarotti, wrote a booklet *Il Newtonianesimo per le dame* ("Newtonianism for ladies") in this regard, which was the first manifesto of the women's liberation movement, translated into most European languages.

Both approaches can be formalized. A familiar example of a formal theory is Euclid's geometry. Once a set of components has been defined and their mutual relations stated, via a group of axioms, all possible consequences are deducible as theorems, which provide by necessity all explanations as well as predictions on the future behavior.

In phenomenology, we have many sciences; in reductionism, we have a single fundamental science, that of the elementary components, out of which we can extract all relevant levels of organization.

Such an approach has been abundantly criticized[12]. The main criticism is that the nonlinear dynamics of microscopic components undergoes multiple bifurcations, of the kind of fig. 2, as a control parameter is varied in order to build up a macroscopic object; thus the construction from scratch of a large system is by no means unique, and the multiple outcomes are a token of *complexity*, as discussed in Sec. 5.

Since this essay however points at a more fundamental approach to our cognitive acts, for the time being we list current approaches without criticism, just to introduce the technical language and get acquainted with the corresponding problems.

---

[12] See Anderson 1972, Arecchi 1992, 1995, Arecchi and Farini 1996.

Reductionism does not mean always to refer to Democritus' atoms (nowadays, we would say to leptons and quarks), but to stop at a suitable level where the elementary components are sufficiently characterized. Such are the biomolecules of living beings.

For all practical purposes, the biologists need the descriptive properties of the biomolecules, plus some knowledge of the nearest lower level, that of atomic physics. Think e.g. of the role of Na^+, K^+ and Ca^{2+} ion conductances in neurophysiology, or of the devastating effect of some atoms as thallium or plutonium on enzymatic processes, or the balance of hydrogen bonds and van der Waals bonds in stabilizing protein folding.

Thus biochemistry is founded on atomic physics, but it does not require nuclear or subnuclear physics. Similarly, atomic physics requires only nuclear physics but not further levels below, and so on.

However, there is no fundamental level which acts as the ultimate explanatory layer. In fact, the problem has recently been addressed whether a formal description of the state of an elementary particle is sufficient to build a faithful replica of it elsewhere (what is termed the teleportation problem[13]). A formal description within the current language of quantum mechanics is not sufficient to provide full recovery of the particle. One must add some non-formalizable information. It is not the place to expand on such a technical aspect, I just recall that the transmitter and receiver must share not only verbal information (the formal description) but they must also be exposed respectively to the two parts of an EPR (Einstein-Podolsky-Rosen) state, or "entangled" pair. By "entangled" we mean a strong quantum correlation which has no classical counterpart, and hence cannot be formalized in the classical language of physics[14]. Just like interacting with a baby or somebody with a different language; nominal definitions are not enough; the dictionary must be integrated by "ostensive definitions", just putting your finger on the object.

We now discuss how the set (1) of relevant variables and the law of motion (2) are established in the two cases *a priori* (or reductionistically) and *a posteriori*, or phenomenologically.

---

[13] See Bennett et al. 1993.

[14] See Bennet et al. 1993.

3.1. A priori

This approach began with Newton, and has continued up to the present search for a unified theory of all fundamental interactions. It consists in counting the particles in the universe, attributing to each one 6 numbers, 3 coordinates in Euclidean space and 3 momenta (or more simply within the non-relativistic limits, 3 velocity components).

Quantum mechanics added more specifications for internal degrees of freedom, such as "spin" and electrical charge, both for leptons and quarks, plus "strangeness", "charm" and some other properties for quarks. The numbers corresponding to these internal degrees of freedom do not span over all real values, but are confined to a small set of possible values. Most often, they correspond to dichotomous variables with just two values, conventionally denoted as 0 and 1. Anyway, each x_i is a group of 6 real numbers for a classical particle, plus a few other discrete numbers for quantum particles.

The coupling function f of Eq. (2) implies mutual relations. Initially, the single universal relation was considered to be Newton's gravitational interactions. Later, Maxwell electromagnetic theory became the prototype of any field theory. Here, the coupling is no longer between particles, but each particle feels forces corresponding to a new entity, the local electromagnetic field at its position. Vice versa, the fields are generated by moving charged particles. Thus the particle-particle interactions are mediated by the fields; in field dynamics no longer we speak of "action at distance".

In electromagnetic theory one adds a new set of x_i consisting of the 6 components of the electric and magnetic field at each point in space. In this case we have a continuous field problem, since the position is not a discrete set of numbers, but varies with continuity. We write $x(r)$ where r denotes the position coordinates in a 3-dimensional space; here r is made of three real numbers which we write as $r \in R^3$ (r belongs to the 3-dimensional real space).

The continuum problem has haunted modern physics since its beginning, and clever devices to deal with it have been produced. However in most cases the continuous fabric of space can be discretized as a lattice of points at finite distances from each other.

I illustrate this trick with reference to a time-dependent signal $x(t)$ observed over a finite time interval T; it depends on all the real values taken by t in the segment T. Outside T the signal is not defined, thus we can arbitrarily assume that it repeats periodically with period T, without affecting the values within the observation interval. This means that its information is contained in a discrete Fourier series of pairs of real numbers (A_n, φ_n) sampled at a frequency $n\dfrac{1}{T}$ which is the n-th harmonic of the fundamental repetition frequency $1/T$, that is

$$x(t) = \sum_{n \cong 1}^{\infty} A_n \cos\left(n\frac{2\pi}{T} + \varphi_n \right) \qquad (7)$$

Thus, the finite interval T limitation has simplified the mathematical description of the signal from continuum to discrete. We do indeed probe with continuity each real t, but we synthesize the signal by summing up at each point a discrete set of sinusoids. If furthermore we consider that any detection or signal processing device is a low pass filter with a finite frequency window B (i.e., it responds only to frequencies up to B), then we can truncate the sum (7) up to a maximum value $n_{max}=B\,T$, and the signal information of $x(t)$ over T is contained in n_{max} sinusoids. Since for each frequency we have an amplitude A_n and a phase φ_n however, the set of numbers which fully specify our signal is twice $B\,T$, that is,

$$N = 2BT \qquad (8)$$

This important sampling theorem, stated by radar investigators during World War II[15], sets the resolution limit for an observation with bandwidth B lasting for a time T. To acquire more information, one must increase either B or T.

In a similar way, a visual system (the eye, or a telecamera) frames a finite two-dimensional domain of sizes L_1, L_2 with bandwidths B_1 and B_2. Thus the number of relevant picture elements (pixels) of a two dimensional image is given by

$$N = 4B_1 B_2 L_1 L_2 \qquad (9)$$

---

[15] See Shannon 1949.

The sampling theorem has induced the strong belief that any cognitive process deals with a finite number of elements, and furthermore that the universe is described by a finite, though very large, number of degrees of freedom.

The mathematics of eighteenth century physics has been expressed in terms of *ODE's* (ordinary differential equations) for the continuous variation of a variable x_i as a continuous time t flows. If infinitesimally close space points have to be coupled, then we express the co-presence of space derivatives together with time derivatives by *PDE's* (partial differential equations). If time is discretized by sampling it at finite distances, then the *ODE's* are replaced by iteration maps, whereby the value of x at the discrete time $n+1$ depends upon the value of x at the previous time n.

If the space can also be discretized as a lattice of disjoint points denoted by discrete indices i,j, then the space derivatives reduce to coupling the iteration maps at different points (*CML* = coupled map lattice).

Eventually, if the variable x is also constrained to assume a finite set of values, in the limit binary values *(0,1)*, then we have a *CAM* (cellular automaton machine) consisting of a network of points each represented by a binary variable which updates at discrete times depending on the values of the neighboring points or "cells"[16]. We summarize in Table 1 the different ways of mathematically modeling the evolution of a physical system.

Table 1

State variable	*Time variable*	*Space variable*	
C	*C*	*C*	*PDE*
C	*C*	*D*	*ODE*
C	*D*	*D*	*CML*
D	*D*	*D*	*CAM*

C = continuous, D = discrete

CAM techniques have been very powerful in dealing with model problems, from biology (genetics, population dynamics, immune systems) to sociology (traffic problems, econophysics) and meteorology. They have

---

[16] See Wolfram 1984.

become the basis of a finitistic ideology, whereby the universe can be seen as a large CAM[17].

A fundamental limitation to this ideology arises however from the quantum non-commutativity of pairs of complementary observables, as we'll discuss in Sec.6.

3.2. A posteriori

New classes of phenomena are disclosed by the exploitation of innovative sophisticated systems of investigation, e.g. recording long time-series in financial trading or in car traffic, imaging techniques in brain investigations, and automatic machines for sequencing DNA. It is very difficult to fit this new phenomena into a Newtonian frame. A component description is hopeless, and one wants to approach the problem directly, without prejudices.

Suppose that, from salience considerations, we have focused our attention on a time-dependent quantity $u(t)$. Salience means that $u(t)$ displays a patterned behavior, that is, it is strongly correlated with its values at later times. Take $u(t)$ as the deviation from an average value, then its time average is zero, that is, $u(t)$ looks like a sequence of +/- values. Consider the product of two u's at different times. If they are unrelated, then the average product of them, called correlation function $C(t,t')$, is also zero.

A nonzero $C(t,t')$ is a signature of salience. Karhunen (1946) and Loeve (1950) introduced independently the following retrieval method that we call KL[18]. If $\theta_n(t)(n=1,2...,L)$ are the L most prominent characteristic functions (called eigenfunctions) which retrieve the correlation $C(t,t')$, and if l is a small number, then we can accurately reconstruct the signal as a weighted sum of L functions as follows:

$$u(t) = \sum_{n=1}^{L} a_n \theta_n(t) \qquad (10)$$

If the signal depends on space rather than time, then we grasp the salient features of a given space pattern. Each of these saliences in general is spread over the whole domain. A relevant example in the convective motion of a

---

[17] See Toffoli 1998.
[18] See Karhunen 1946, Loeve 1955.

fluid was given by Ciliberto et al. 1991, where the three main "modes" of behavior ($L=3$) are distributed over the whole fluid cell.

The opposite limit occurs when saliences are strongly localized. Think e.g. of the face elements (nose, eyes, mouth shape) upon which identikits of criminals are built in police investigations. In such a case, KL would be inconvenient, since it requires a large L to converge toward a localized feature. Here the successful phenomenological approach is just the opposite of KL. It consists in reconstructing a pattern, e.g. a face, by a small series of "prototypes". This approach is used in many machine vision programs[19].

4. Closed versus open systems

I have discussed elsewhere[20] the failure of what Anderson called the "constructionist" program[21]. Trying to build a structured system out of its elementary components does not provide a univocal outcome.

Indeed the components interact via a nonlinear law as Eq.(2), and the emergence of a new stable structure starting from an initial condition P_0 requires the appropriate tuning of the control parameters. Such a tuning provides in general more than one new stable state. (See fig.3).

The emerging states $2,2'$ are equivalent, thus, as μ is tuned to μ_{C1}, the system has equal probability of emerging in the state 2 or $2'$, unless we break the symmetry of the bifurcation by the application of an external field (see fig.3b), which makes the two stable states non-equivalent, and hence one of the two (the upper one in the figure) chosen with higher probability. The number of equivalent outcomes increases exponentially with the order of the bifurcation: 2 at μ_{C1}, 4 at μ_{C2} and so on.

Hence, a reductionistic program based on the dynamical description of the components does not provide a unique outcome. We must assign some extra information consisting of possible external fields, which univocally specify the final state.

But external fields are beyond the information provided by the dynamic properties of the components.

---

[19] See Weber et al. 2000.
[20] See Arecchi 1995.
[21] See Anderson 1972.

We must then distinguish between *closed* and *open* systems. The first are surrounded by walls which provide precise boundary conditions. Their evolution yields multiple outcomes so that we predict "potentialities", never the "actual" system that is observed.

In the case of open systems we must augment our description including the values of the external fields which select a unique outcome at each bifurcation. In general we don't know how to do that a priori; we can rather proceed backward to an historical reconstruction of external contributions which have obliged our open system to evolve from an initial to a final state. Notice that the setting of μ on the horizontal axis is at our will for a closed system, whereas in an open system it has a proper evolution in time that we do not control. Thus a bifurcation tree such as fig. 3 looks like an evolutionary tree, usually plotted in biology as a function of time.

We might think that a metalevel description could treat the overall situation (set of components plus external process) as a closed system. In fact, we would transfer the ambiguities at the metalevel, and to recover uniqueness the metalevel must be affected by its own external forces and so on. In other words, by treating successive layers we are dealing with an "onion" structure.

A global description of the whole cosmic onion as a closed system is the dream of theoretical physics. A TOE (Theory Of Everything) would be an equation like (2) where $\underset{\approx}{x}$ are now all the degrees of freedom of the universe and $\underset{\approx}{f}$ is the unified mathematical formulation that one day will be reached among electro-weak, strong and gravitational interactions. In such a situation there would be *nothing* left out, thus "new" must itself be a function of $\underset{\approx}{x}$, and hence Eq. (2) of TOE would be *closed*, with no external control parameters.

In fact, this is in principle impossible. Any foreseeable nominal description (that is, expressed by precise numbers) is incomplete even for a single particle. Therefore we must split the vector x_{tot} of all the degrees of freedom of the universe into an observable set x_0 and a complementary set $\bar{x}$ which escapes our description. Our relevant physical equation refers only to the observed part x_0. Thus Eq. (2) must be re-written as

$$\dot{x}_0 = \underset{\approx}{f}(x_0, \mu(x_{0'}, \bar{x})). \qquad (11)$$

This form of expression shows that the $\bar{x}$ dependence of μ excludes the above equation from being closed.

5. Complication versus complexity

5.1. Complexity of symbolic sequences

In computer science, we define the complexity of a word (symbol sequence) as some indicator of the cost implied in generating that sequence[22]. There is a "space" cost (length of the instruction stored in the computer memory) and a "time" cost (the CPU time for generating the final result out of some initial instruction).

A space complexity called AIC (Algorithmic Information Complexity)[23] is defined as the length in bits of the minimal instruction which generates the desired sequence. This indicator is *maximum* for a random number, since there is no compressed algorithm (that is, shorter than the number itself) to construct a random number.

A time complexity called "logical depth"[24] is defined as the CPU time required to generate the sequence out of the *minimal* instruction. It is minimal for a random number, indeed, once the instruction has stored all the digits; just command: PRINT IT.

Of course, for simple dynamical systems such as a pendulum or the Newtonian two-body problem, both complexities are minimal.

While AIC refers to the process of building a single item, logical depth corresponds to finding the properties of all possible outputs from a known source.

In fact, the exact specification of the final outcome is beyond the ambition of the natural scientist, whose goal is more modest. It may be condensed to the two following items:

---

[22] See Hopcroft and Ullman 1979.

[23] See Kolmogorov 1965, Chaitin 1966.

[24] See Bennett 1987.

i) to transmit some information, coded in a symbol sequence, to a receiver, possibly economizing with respect to the actual string length, that is, making good use of the redundancies (this requires a preliminary study of the language style);

ii) to predict a given span of the future, that is to assign with some likelihood a group of forthcoming symbols.

For this second goal, introduction of a probability measure is crucial[25] in order to design a complexity-machine, able to make the best informational use of a given data set.

Such a machine which should mimic the scientific procedure acts as follows[26]. Assume that a group of measuring apparatuses have provided the agent A with information coded as a numerical sequence s (for convenience we use a binary code, so that the length $|s|$ of s is measured in bits). Agent A has a good understanding of what happens if it can transfer to a received B compact information y upon which B can reconstruct the sequence $s'=s$. Of course, y has to be shorter than s otherwise it would be a tautology, which implies no understanding whatsoever. Thus A is obliged to recur to a class of models built in its memory. Suppose it has chosen a model m; then A can simulate the behavior of the observed system and realize that there is an error e between the actual measurement s and its model reconstruction. If B receives both information m and e, then B can reconstruct $s'=s$. The bit length of the transmitted information is

$$|y| = |m| + |e| \qquad (12)$$

and it has to be minimized for a successful description.

In this case we call the complexity of the explanation the compression ratio

$$C(m,s) = \frac{|y|}{|s|} = \frac{|m| + |e|}{|s|} \qquad (13)$$

The value of C is bounded above by 1; it depends upon the choice of the model m. There are two limit cases for which $C=1$ is the worst. When the

---

[25] See Grassberger 1986, Gell-Mann 1994.

[26] See Crutchfield 1992.

model is trivial ($|m| = 0$) the entire information is on the error channel ($|e| = |s|$). When the model is tautological $m=s$ there is no error and $|e| = 0$.

The class of models can be scanned by a Bayes rule[27]. This is the case of an "expert system" equipped with a class of models, within which the system formulates the best diagnosis by minimizing C.

5.2. *A dynamic approach to complexity*

As discussed in Sec. 3, the reductionist approach consists of building a hierarchy from large to small and showing how the behavior of smaller objects should determine that of larger ones. But here a perverse thing occurs. If our words were a global description of the object in *any* situation (as the philosophical "essences" in Galileo) then, of course, knowledge of elementary particles would be sufficient to make predictions on animals and society. In fact, Galileo's self-limitation to some "affections" is sufficient for a limited description of the event, but only from a narrow point of view. Even though we believe that humans are made of atoms, the affections that we measure in atomic physics are insufficient to make predictions of human behavior.

We call *complexity* the fact that higher levels of organization display features not predictable from the lower ones, as opposed to the previous computer cost of a symbolic task, which we rather call *complication* .

In this way, complexity is not a property of things (like being red or hot) but is a relation to the status of our knowledge, and for modern science it emerges from Galileo's self-limitation.

Reductionism from large to small was accompanied by a logical reduction of the scientific explanation to a deductive task from a set of axioms.

In this spirit, a scientific theory is considered to be a set of primitive concepts (defined by suitable measuring apparatuses) plus their mutual relations. Concepts and relations are the axioms of the theory. The deduction of all possible consequences (theorems) provides predictions which have to be compared with the observations. If the observations falsify the expectations, then one tries with different axioms.

The deductive process is affected by Gödel undecidability as is any formal theory, in the sense that it is possible to build a well-formed

---

[27] See Crutchfield et al. 1989.

statement, but the rules of deduction are unable to decide whether that statement is true or false.

Apart from this, a second drawback is represented by *intractability*, that is, by the exponential increase of possible outcomes among which we have to select the final state of a dynamical evolution. As discussed in Sec. 4, during the dynamical evolution of an open system the control parameters $\{\alpha\}$ may assume different values, hence the cascade of bifurcations provides a large number of final states starting from a unique initial condition.

Thus the reductionist suggestion of explaining reality out of its constituents yields an exponentially high number of possible outcomes, when only one is in fact observed. This means that, while the theory, that is the syntax, would give equal probability to all branches of the tree, in reality we observe an *organization process*, whereby only one final state has a high probability of occurrence.

Hence, whenever we are in the presence of *organization*, that is of a unique final state, this means that at each bifurcation vertex the symmetry has been broken by an external agent which forces a unique outcome, as shown in Fig. 3.

We can thus summarize the logical construction (to rephrase Carnap 1967) of a large system out of its components as follows:

i) A set of control parameters is responsible for successive bifurcations leading to an exponentially high number of final outcomes. If the system is "closed" to outside disturbances, then all outcomes have comparable probabilities, and we call complexity the impossibility of predicting which one is the state we are going to observe.

ii) For a system of finite size embedded in an environment, a set of external forces is applied at each bifurcation point, which break the symmetries, biasing toward a specific choice and eventually leading to a unique final state.

We are in the presence of a conflict between (i) "syntax" represented by the set of rules (axioms) and (ii) "semantics" represented by the intervening external agents. The syntax provides many possible outcomes. But if the system is open, then it organizes to a unique final outcome. Once the syntax is known, the final result is therefore an acknowledgement that the set of external events must have occurred, that they have made the evolution meaningful (whence "semantics").

We define "certitude" as correct application of the rules, and "truth" the adaptation to the reality. However, due to the freedom we have in formulating theoretical conjectures, the same final outcome could be reached by a different set of rules, corresponding to a different syntactic tree. In such a case, retracing the new tree of bifurcations, we would reconstruct a different set of external agents. Thus, it seems that truth is language dependent!

Furthermore, this freedom in choosing the rules (the syntax) means that we can even find a set of axioms which succeeds in predicting the correct final state without external perturbations. This is indeed the pretension of what is termed "autopoiesis", or "self-organization"[28].

From a cognitive point of view, a self-organized theory can thought of as a "petitio principi", a tricky formulation tailored for a specific purpose and not applicable to different situations. Rather than explicitly detailing the elements of the environment which break the symmetry, the supporter of the self-organized theory has already exploited at a pre-formalized level a detailed knowledge of the process in planning appropriate axioms.

An "ad-hoc" model may fit a specific situation, but in general it lacks sufficient breadth to be considered as a general theory. Think e.g. of the Ptolemaic model of the solar system, which holds only for an Earth-based observer, as compared to the Newtonian one, which also holds for an observer travelling through the solar system.

However, in describing the adaptive strategy of a living species, or a community etc., "self-organization" may be the most successful description. In other words, once the environmental influences are known, it is better to incorporate their knowledge into the model, thus assuring the fast convergence to a given goal.

5.3. Complexity differs from complication

When all the rules of a game and all the partners (components) have been introduced, we are in the presence of a definite symbolic system. The corresponding problems can be solved at a cost which may increase more than polynomially with the number of partners, e.g. consider the Travelling Salesman problem or TSP. We prefer to call "complication" the difficulty of

---

[28] See Krohn et al. 1990.

solving a problem within a formal system, and use "complexity" to denote any cognitive task before an open system. In such a case, a cognitive machine as an "expert system" is limited to a finite set of models m. Furthermore, it is bound to a precise setting of the measuring apparatuses: take for instance the apple properties listed in Sec. 2. As discussed in Arecchi 2001 this limitation of an expert system is overcome by an adaptive strategy, whereby an agent spans not only the class of available models by a Bayesian strategy, as in the computer-based model-reasoning peculiar to expert systems[29], but also changes in the course of time the type of measures performed, thus reaching a meta-level where the cognitive agent is equivalent to a large class of expert systems. Note that "large" can be still finite. In Sec. 6 we will discuss a fundamental quantum relation for time dependent processes, which exclude finitism.

6. The neurophysics of perception

6.1. What is neurophysics?

It is by now firmly established that a holistic perception emerges out of separate stimuli entering different receptive fields, by synchronizing the corresponding spike trains of neural action potentials[30].

We recall that action potentials play a crucial role in communication between neurons[31]. They are steep variations in the electric potential across a cell's membrane, and they propagate in essentially constant shape from the soma (neuron's body) along axons toward synaptic connections with other neurons. At the synapses they release an amount of neurotransmitter molecules depending upon the temporal sequences of spikes, thus transforming the electrical into a chemical carrier.

As a matter of fact, neural communication is based on a temporal code whereby different cortical areas which have to contribute to the same percept P synchronize their spikes. Spike emission from a nonlinear-threshold dynamical system results as a trade-off between bottom-up stimuli to the higher cortical regions (arriving through the LGN (Lateral Geniculate

---

[29] See Magnani et al. 1999.
[30] See Von der Malsburg 1981, Singer and Gray 1995.
[31] See Izhikevich 2000.

Nucleus) from the sensory detectors, video or audio) and threshold modulation due to top-down readjustment.

It is then plausible to hypothesize, as in ART (Adaptive Resonance Theory[32]) or other computational models of perception[33] that a stable cortical pattern is the result of a Darwinian competition among different percepts with different strength. The winning pattern must be confirmed by some matching procedures between bottom-up and top-down signals.

We present two fundamentals aspects of percept formation, namely,

- *The neurodynamics of spike formation.*
- *A quantum limitation in information encoding/decoding through spike train.*

As for the first aspect, a saddle-point instability separates in parameter space an *excitable* region, where axons are silent, from a *periodic* region, where the spike train is periodic (equal interspike intervals). If a control parameter is tuned to the saddle point, the corresponding dynamic behavior (homoclinic chaos) consists of a frequent return to instability[34]. This manifests as a train of geometrical identical spikes, which however occur at erratic times (chaotic interspike intervals). Around the saddle point the system displays great susceptibility to external stimulation, hence it is easily adjustable and prone to respond to an input, provided this is at sufficiently low frequencies; this means that such a system is robust vis-à-vis broadband noise, as will be discussed later.

As for the second aspect, temporal coding requires a sufficiently long sequence of synchronized spikes in order to realize a specific percept. If the sequence is interrupted by the arrival of new uncorrelated stimuli, then a fundamental uncertainty ΔP emerges in the percept space P. This is related to the finite duration ΔT allotted for the code-processing by a fundamental uncertainty relation

$$\Delta P \cdot \Delta T \geq C,$$

where C is a positive dimensional quantity whose non-zero value represents a quantum constraint on the coding. This constraint implies that the percepts are not set-theoretical objects, that is, objects belonging to separate domains, but rather that there exist overlap regions where it is

---

[32] See Grossberg 1955a and 1955b.

[33] See Edelman and Tononi 1995.

[34] See Allaria et al. 2001.

impossible to discriminate one percept from another. We will discuss the occurrence of this new class of time-dependent perceptual illusions later.

We call "neurophysics" the combination of i) and ii), in analogy with "econophysics" which has extracted some general physical phenomena from economic phenomena[35].

Neurophysics is distinct from neurodynamics, which is the investigation of dynamic models of neuron behavior, as well from neurophysiology, which explores the coupling of different brain areas. Neurophysics is restricted to the two above items, and it is rather model-independent, so that it provides a general ground upon which different models can be constructed and compared.

6.2. The role of duration T in perceptual definitions: a quantum aspect

How does a synchronized pattern of neuronal action potentials become a relevant perception? This is an area of active investigation which may be split into many hierarchical levels. At the present level of knowledge, we think that not only the different receptive fields of the visual system, but also other sensory channels such as the auditory, olfactory, etc. integrate via features binding them into a holistic perception. Its meaning is "decided" in the PFC (PreFrontal Cortex), which is a kind of arrival station from the sensory areas and departure point for signals going to the motor areas. On the basis of the perceived information, actions are commenced, including linguistic utterances.

Sticking to the neurodynamical level, which is the most fundamental, and leaving to other sciences, from neurophysiology to psychophysics, the investigation of what goes on at higher levels of organization, we stress here a fundamental temporal limitation. Taking into account that each spike lasts about 1 msec, that the minimal interspike separation is 3 msec, and that the average decision time at the PFC level is about T=240 msec, we can split T into 240/3 =80 bins of 3 msec duration, which are designated by 1 or 0 depending whether they have a spike or not. Thus the total number of messages which can be transmitted is

$$2^{80} \approx 10^{27}$$

that is, well beyond the information capacity of present computers. Even though this number is large, we are still within a finitistic realm. Provided we have time enough to ascertain which one of the 10^{27} different messages

---

[35] See Mantegna and Stanley 2000.

we are dealing with, we can classify it with the accuracy of a digital processor, without residual error.

But suppose we expose the cognitive agent to fast-changing scenes, for instance by presenting unrelated video frames with a time separation less than 240 msec. While small gradual changes induce the sense of motion as in movies, large differences imply completely different subsequent spike trains. Here any spike train is interrupted after a duration ΔT less than the canonical T. This means that the PFC cannot decide among *all* perceptions having the same structure up to ΔT, but different afterwards, that are coded by the neural systems. How many are they: the remaining time is $\tau=T-\Delta T$. To give a numerical example, take a time-separation of the video frames $\Delta T=T/2$, then $\tau=T/2$. Thus in spike space an interval ΔP comprising

$$2^{\tau/3} \approx 2^{40} \approx 10^{13}$$

different perceptual patterns is uncertain. As we increase ΔT, ΔP reduces, thus we have an uncertainty principle

$$\Delta P \cdot \Delta T \geq C.$$

The problem faced thus far in the scientific literature, of an abstract comparison of two spike trains without accounting for the available time for such a comparison, is rather unrealistic. A finite available time ΔT plays a crucial role in any decision, whether we are trying to identify an object within a fast sequence of different perceptions, or scanning through memorized patterns in order to decide about an action.

As a result, the perceptual space P per se is meaningless. What is relevant for cognition is the joint space (P,T), since "in vivo" we have always to face a limited time ΔT which may truncate the whole spike sequence upon which a given perception has been coded. Only "in vitro" we allot to each perception all the time necessary to classify it.

A limited ΔT is not only due to the temporal crowding of sequential images, as reported clinically in behavioral disturbances in teenagers exposed to fast video games, but also to sequential conjectures that the semantic memory essays via different top-down signals. Thus in the metrical space (P,T), while the isolated localization of a percept P (however long T is) or of a time T (however spread the perceptual interval ΔP is) have a sense, a joint localization both in percept and time has an ultimate limit when the corresponding domain is less than the quantum area C.

Let us consider the following thought experiment. Take two percepts P_1 and P_2 which for long processing times appear as the two stable states of a bistable optical illusion, e.g. the Necker cube. If we allow only a limited observation time ΔT, then the two uncertainty areas overlap. The contours drawn in fig. 4 have only a qualitative meaning. The situation is logically equivalent to the non-commutative coordinate-momentum space of a single quantum particle. In this case it is well known[36] that the quasiprobability Wigner function has strong non-classical oscillations in the overlap region. We cannot split the coordinate-momentum space into two disjoint domains (sets) to which we can apply a Boolean logic or a classical Kolmogorov probability. This is the reason why the Bell inequalities are violated in an experiment dealing with such a situation[37]. The Wigner function formalism derives from a Schroedinger wavefunction treatment for pure state, and corresponding density matrix for mixed states.

In the perceptual (P,T) space no Schroedinger treatment is yet available, but we can apply a reverse logical path as follows.

The uncertainty relation $\Delta P \cdot \Delta T \geq C$ forbids a partition of the (P,T) space into sets. Therefore the (P,T) space is non commutative. Thus it must be susceptible of a Wigner function treatment, and we can consider the contours of fig. 4 as fully equivalent to isolevel cuts of a Wigner function. Hence we can introduce Schroedinger cat states and violations of Bell inequalities exactly as in quantum physics but with a reverse logical process, as illustrated in fig.5.

The equivalent of a superposition state should be a bistable situation observed for a time shorter than the whole decision time. An experimental test is in preparation in my research group. Such a test should provide an estimation of the C value, which plausibly changes from individual to individual, and for a single one may be age- and motivation-dependent.

Thus in neurophysics time occurs in two completely different senses, that is, as the ordering parameter to classify the position of successive events, and as the useful duration of a relevant spike sequence, that is, as the duration of a synchronized train. In the second meaning, time T is a variable conjugate to perception P.

The quantum character has emerged as a necessity from the analysis of an interrupted spike train in a perceptual process. It follows that the (P,T) space cannot be partitioned into disjoint sets to which a Boolean yes/no relation is applicable, and hence where ensembles obeying a classical probability can be considered. A set-theoretical partition is the condition for applying the Church-Turing thesis, which establishes the equivalence

---

[36] See Zurek 1991.

[37] See Omnès 1994.

between recursive functions on a set and operations of an universal computer machine.

The evidence of the quantum entanglement of overlapping perception should rule out in principle the perceptual processes' having a finitistic character. This should be the negative answer to the 1950 Turing question whether mental processes can be simulated by a universal computer[38]. Among other things, the characterization of the "concept" or "category" as the limit of a recursive operation on a sequence of individual related perceptions gets rather shaky, since recursive relations imply a set structure.

Quantum limitations were also put forward by Penrose 1994 but on a completely different basis. In his proposal, the quantum character was attributed to the physical behavior of the "microtubules", which are microscopic components of the neurons that play a central role in synaptic activity. However, speaking of quantum coherence in biological processes is very hard to accept, if one accounts for the extreme vulnerability of any quantum system as being due to "decoherence" processes, which make quantum superposition effects observable only in extremely controlled laboratory situations.

References

Agazzi, E.: 1974, *Temi e Problemi di Filosofia della Fisica*. Roma: Edizioni Abete.
Allaria, E., Arecchi, F.T., Di Garbo, A. and Meucci, R.: 2001, "Synchronization of homoclinic chaos", *Phys. Rev. Lett* .86, 791.
Anderson, P. W.: 1972, "More is different", *Science* 177, pp. 393-396.
Arecchi, F. T.: 1992, "Models and metaphors in science", in B.Pulmann (ed.): 1992, *Proceedings of Pontifical Academy of Sciences, Plenary Session on 'Emergence of Complexity in Mathematics, Physics, Chemistry and Biology', Rome 26-31 Oct. 1992*. Princeton NJ: Princeton University Press.
— 1995, "Truth and certitude in the scientific language", in F. Schweitzer, Gordon and Breach (ed.): 1997, *Self-Organization of Complex Structures from Individual to Collective Dynamics*. Amsterdam.
— 2000, "Complexity and adaptation: a strategy common to scientific modeling and perception", *Cognitive Processing* 1, 23.
— 2001, "Complexity versus complex system: a new approach to scientific discovery", *Nonlin. Dynamics, Psychology, and Life Sciences*, 5, 21.
Arecchi, F. T. and Farini, A.: 1996, *Lexicon of Complexity*. Firenze: Studio Editoriale Fiorentino.

---

[38] See Turing 1950.

Bennett, C. H.: 1987, "Dissipation, information, computational complexity and the definition of organization" in D. Pines (ed.), *Emerging Syntheses in Science.* Redwood City, Ca: Addison Wesley.

Bennett, C. H., Brassard, G., Crepeau, C., Jozsa, R., Peres, A., and Wootters, W.: 1993, "Teleporting an Unknown Quantum State via Dual Classical and EPR Channels", *Phys. Rev. Lett.* 70, pp. 1895-1899.

Capra, F.: 1975, *The Tao of Physics: an Exploration of thePparallels between Modern Physics and Eastern Mysticism"*. Washington D.C.: The International Bestseller (3^d ed. 1991).

Carnap, R.: 1967, *The Logical Construction of the World*. Berkely, CA: University of California Press.

Chaitin, G. J.: 1966, "On the length of programs for computing binary sequences", *J. Assoc. Comp. Math.* 13, pp. 547-560.

Churchland, P. S. and Sejnowski T. J.: 1992, *The Computational Brain*. Cambridge, Mass: MIT Press.

Ciliberto, S. and Nicolaenko B.: 1991, "Estimating the number of degrees of freedom in spatially extended systems", *Europhys. Lett.* 14, 303.

Crutchfield, J. P. and Young, K.: 1989, "Inferring statistical complexity", *Phys. Rev. Lett* 63, 105.

Edelman, G.M. and Tononi, G.: 1995, "Neural Darwinism: The brain as a selectional system", in J. Cornwell, (ed.), *Nature's Imagination: The Frontiers of Scientific Vision*. New York: Oxford University Press, pp. 78-100.

Evans, R.I.: 1973, *Jean Piaget: the Man and His Ideas*. New York: E.P. Dutton.

Feyerabend, P.: 1975, *Against Method*. London: Verso.

Galilei, G.: 1932, *Letter to M. Welser "On the sun spots", 1 December 1612*. Italian original in: *Opere di G. Galileo,* Vol. V. Firenze, pp. 187-188.

Gell-Mann, M.: 1994, *The Quark and the Jaguar*. New York: W.H. Freeman.

Grassberger, P.: 1986, "Toward a quantitative theory of self-generated complexity", *Int. J. Theor. Phys.*, 25, pp. 907-919.

Grossberg, S.: 1995a, "The attentive brain", *The American Scientist*, 83, 439.

— 1995b, "Review of the book by F. Crick: *The Astonishing Hypothesis: The Scientific Search for a Soul*", 83 (n.1).

Guarino, N.: www.ladseb.pd.cnr.it/infor/Ontology/ontology.html.

Haken, H.: 1983, *Synergetics, An Introduction*. Berlin: Springer Verlag (3rd edition).

Hopcroft, J. E. and Ullman, J. D.: 1979, *Introduction to Automata Theory, Languages and Computation*. Reading, MA: Addison-Wesley.

Hubel, D. H.: 1995, *Eye, Brain and Vision*, Scientific American Library, no. 22. New York: W.H. Freeman.

Izhikevich, E.M.: 2000, "Neural Excitability, Spiking, and Bursting", *Int. J. of Bifurcation and Chaos* 10, 1171.

Karhunen, K.: 1946, "Zur Spektraltheorie Stochasticher Prozess", *Ann. Acad. Sci. Fennicae* 37.

Kolmogorov, A.N.: 1965, "Three approaches to the quantitative definition of information", *Problems of Information Trasmission* 1. pp. 4-20.

Krohn, W., Küppers, G. and Nowotny, H.: 1990, *Self-organization: Portrait of a Scientific Revolution*. Dordrecht: Kluwer Academic Publishers.

Landau, L. D. and Lifshitz, E.M.: 1980, *Statistical Physics*. Pergamon: New York (3rd edition).

Loève, M. M.: 1955, *Probability Theory*. Princeton, N.J.: VanNostrand.

Magnani, L., Nersessian, N. J. and Thagard, P. (eds.): 1999, *Model-based Reasoning In Scientific Discovery*. New York: Kluwer.

Mantegna, R. N. and Stanley, H. E.: 2000, *An Introduction to Econophysics*. Cambridge UK: Cambridge University Press.

Omnès, R.: 1994, *The interpretation of Quantum Mechanics*. Princeton NJ: Princeton University Press.

Penrose, R.: 1994, *Shadows of the Mind*. New York: Oxford University Press.

Poli, R.: www.formalontology.it/polir.htm

Rieke, F., Warland, D., de Ruyter van Steveninck R. and Bialek W.: 1996, *Spikes: Exploring the Neural Code*". Cambridge Mass: MIT Press.

Rodriguez, E., George, N., Lachaux, J. P., Martinerie J., Renault B.and Varela F.: 1999, "Perception's shadow: Long-distance synchronization in the human brain", *Nature* 397, pp. 340-343.

Shannon, C.E.: 1949, "Communication in the Presence of Noise", *Proc. IRE*, 37, 1.

Simon, H.: 1982, "The architecture of complexity", *Proc. Amer. Philos. Soc.*, 106, pp. 467-485.

Singer, W. E, Gray, C. M.: 1995, "Visual feature integration and the temporal correlation hypothesis", *Ann. Rev. Neurosci.* 18, 555.

Smith, B.: http://wings.buffalo.edu/philosophy/faculty/smith/.

Thom, R.: 1975, *Structural stability and morphogenesis*. Reading, Mass.: Benjamin.

— 1988, *Esquisse d'une sémiophysique*. Paris: InterEditions.

Toffoli, T.: 1998, "Non -conventional computers", in J. Webster (ed.), *Enc.of El. and Electronic Eng.* J. Wiley and Sons.

Turing, A.: 1950, "Computing Machinery and Intelligence", *Mind* 59, 433.

Von der Malsburg, C.: 1981, "The correlation theory of brain function", reprinted in E. Domani, J.L. Van Hemmen and K. Schulten (eds.), *Models of Neural Networks II*. Berlin: Springer.

Weber, M., Welling M and Perona P.: 2000, "Unsupervised Learning of Models for Recognition", in *Proc. 6th Europ. Conf. Comp. Vis., ECCV2000* (Dublin, Ireland, June 2000).

Wolfram, S.: 1984, "Cellular Automata as Models of Complexity", *Nature*, 311, pp. 419-424.

Zurek, W. H.: 1991, "Decoherence and the transition from quantum to classical", *Phys. Today*, (October issue), p. 36.

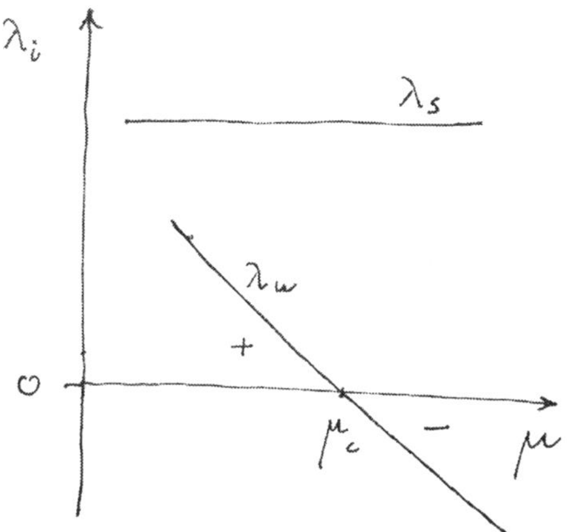

Fig. 1: When the control parameter crosses the value μ_c, the eigenvalues λ_s remain positive, whereas the eigenvalue λ_u goes from positive to negative.

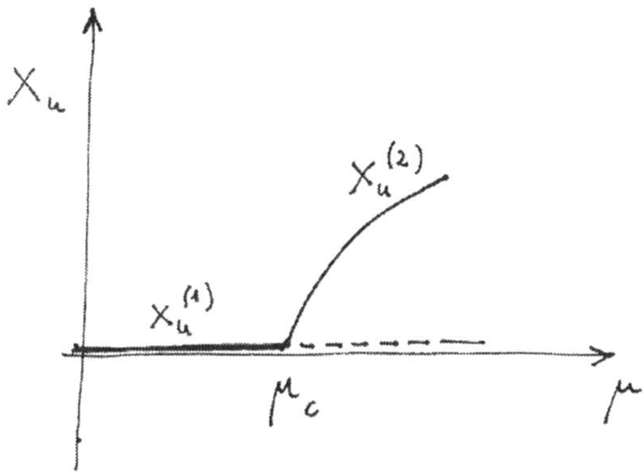

Fig. 2: The horizontal branch $x_u^{(1)}$ becomes unstable at μ_c, and a new stable branch $x_u^{(2)}$ emerges from the bifurcation point.

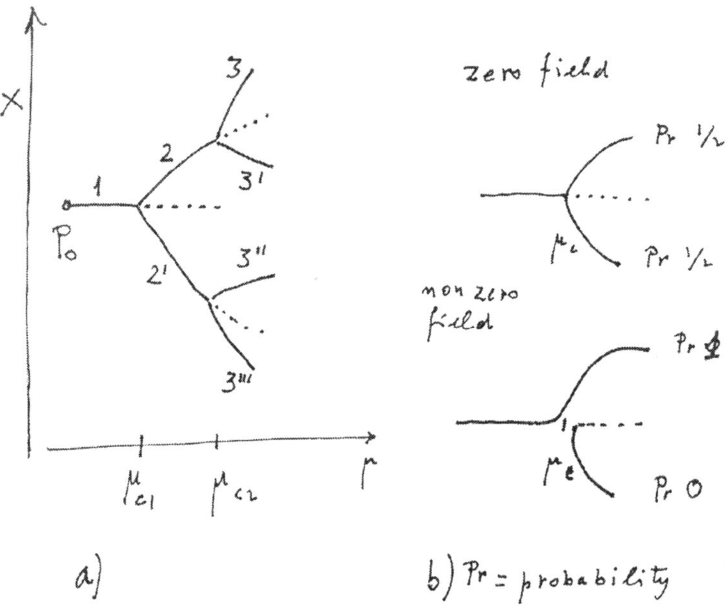

Fig. 3 (a): Example of bifurcation diagram. The dynamical variable x (order parameter) varies vertically, the control parameter μ varies horizontally. Solid (dashed) lines represent stable (unstable) steady states as the control parameter is changed.

(b) Upper: symmetric bifurcation with equal probabilities for the two stable branches.

Lower: asymmetric bifurcation in the presence of an external field. If the gap introduced by the field between the upper and lower branches is wider than the range of thermal fluctuations at the transition point, then the upper (lower) branch has probability 1 (0).

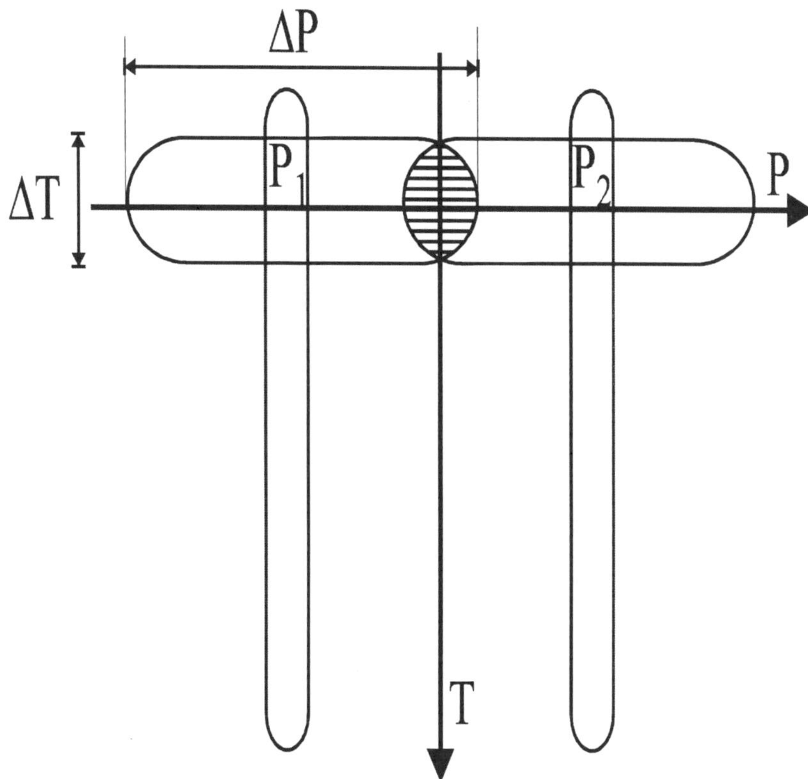

Fig.4: Uncertainty areas of two perceptions P_1 and P_2 for two different durations of the spike trains. In the case of short ΔT, the overlap region is represented by a Wigner function with strong positive and negative oscillations which move as $\cos\dfrac{\Delta P}{C}T$ along the T axis, and thus with a frequency given by the ratio of the percept separation $\Delta P = P_2 - P_1$ to the perceptual "Planck's constant" C.

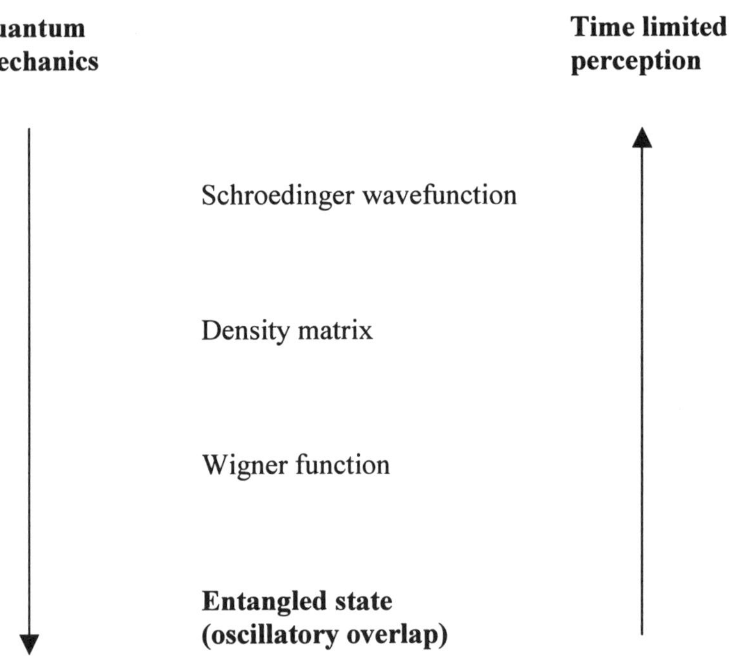

Fig.5: Direction of the logical processes which lead from wavefunction to entangled states or viceversa.

11. Complexity and the Emergence of Intentionality: Some Misconceptions

Mario Casartelli

Dept. of Physics, University of Parma, Italy

Section 1

The thesis I would like to critically focus on may be summarized in the following simple statement: intentionality, as a distinctive feature of conscious mental processes, is an epiphenomenon that emerges out of the complexity of the brain's structure and activity. This idea is a meaningful constituent of a more general program, the reduction of mentalism to brain operations, which, as its supporters are aware, is the materialist and rationalist solution to the mind-body problem. Such a reductionist thesis is obviously related to similar theses regarding the nature of consciousness, free will, qualia, etc., and has several non-uniform facets or branches. We can find it, typically, in the Strong Artificial Intelligence (SAI) program, as well as in severe critics of SAI, such as John Searle, in certain functionalists, and, more generally, in all those who, rejecting the (Cartesian) dualism of *res cogitans* and *res extensa* as well as the idealistic monism of "spirit", feel it a duty to embrace a materialistic monism.

I will not insist here on Cartesian dualism's being the only way to oppose materialism. On the contrary, I will methodologically accept these premises, in order to check their meaning and internal consistency through some of their implications.

Section 2

I must declare in advance in which sense I will use such terms as *emergence, epiphenomenon*, and *complexity*.

Emergence here means not simply that a phenomenon or a feature (such as intentionality) cannot be conceived without a material background, but, more, that the increasing complexity of this background is its sufficient *raison d'être*, without any further extrinsic cause. I shall occasionally use the term *embodiment* as opposite or complementary to *emergence*. Thus, *emergence* refers typically to a property or behavior of a collective in itself, without the input of information from outside (apart from, possibly, the tuning of some parameters); *embodiment* refers, on the other hand, to a property or behavior which, even if realized by a collective, has its formal reason elsewhere. For definiteness: liquidity, as a collective property of H_2O molecules, is a genuine emergent feature. But music is not an emerging property of an orchestra. It would be more correct to say that music is occasionally embodied in an orchestral set.

An *epiphenomenon* is a secondary feature of something more general and fundamental. So, for instance, a traffic jam is an epiphenomenon of holidays, not the contrary.

As to *complexity*, as often noted, it is a term of common language, whose formal characterization has become a matter of research in mathematics, physics, and information and systems theory, especially after the works of Kolmogorov, Chaitin and Salomonov in the sixties. Nowadays, it is a pervasive notion that is turned to in all circumstances where a "third way" between simple order and bare chaos is required[1].

Consider for instance a geometrical context: a random distribution of black and white pixels becomes a uniform gray color when it is seen from far enough away; a smooth pattern, on the contrary, becomes more and more simple when its details are enlarged. But there exists a feature, scale invariance, alternative to both these: this means that once a pattern appears in an "observation window" with respect to a parameter, the same pattern is reproduced, at least qualitatively, by changing the window. "Zooming in" in one sense or the other cannot simplify anything. If the window parameter is spacelike, one obtains fractality, if it is timelike one obtains colored noise, etc.

---

[1] See Li and Vitànyi 1997.

Other characterizations of complexity stress the richness of the information needed to describe a pattern, or to produce it. Still others include the difficulty in performing a particular task. Others the ratio between the length of a program and the length of its output. Others a program's incompressibility. There are relations among these complexities, but not, in general, a complete overlap. One could say that the choice of one or another complexity is already an indication of the observer's intentionality!

A common feature of all these definitions (or, better, a step preliminary to the definitions) is the immersion of the general problem in a formalized context, where such concepts as observables, parameters, configuration spaces, probability measures, languages, purposes, etc. have a definite meaning. Only after such preliminaries, which are far from neutral, can definitions be given.

For instance, if the information requested to describe a pattern refers to the exact position of every point, and if the complexity is defined as difficulty in compressing information, then the maximum of complexity is attained by a random set, because its description cannot be compressed. But if only qualitative features of correlations and the probability distribution of points are considered, then the random set is relatively simple, and other more structured patterns, e.g. fractals, are more complex.

Section 3

Which of these complexities is involved in discussions regarding the emergence of intentionality? We could say all of them, but none of them exclusively. Very often one deals with the old, everyday-language idea of complexity as opposite to simplicity; sometimes, we find an idea of complexity related to unpredictability or chaos; or else to a great number of degrees of freedom; sometimes, when computability, or recursive function theory, or the relevance of quantum effects are referred to, algorithmic complexity is also evoked[2]. But there is no uniform or systematic usage. Not even the widespread idea that simple engines cannot exhibit intentionality is universal (see e.g. J. McCarthy[3] for the opposite opinion). This makes the

---

[2] See Penrose 1994, Grush and Churchland P.S. 1995, Penrose and Hameroff 1995.

[3] See McCarthy 1983. See also http://www.formal.stanford.edu/jmc/

discussion more difficult. I shall proceed therefore by touching only on some typical items.

Section 4

Strong Artificial Intelligence has the advantage of providing a definite model for the discussion. SAI claims a coincidence between thinking and computing, and this claim, in the spirit of "identity theory", should be checked only in terms of computer behavior. Identity theory is the (differently stated but widely shared) idea that what a machine does, in terms of performance, coincides with its essence as a cognitive system[4]: criticism of Searle's Chinese Room Argument, for instance, essentially points to this.

Incidentally, a sort of identity theory (possibly expressed in different ways) lies behind other approaches, which equally insist on complexity— perhaps that this is a way to exorcise the dreaded presence of the famous non-physical "homunculus"[5]. Besides SAI, this exigency appears for instance in eliminative materialism and in neurophysiological approaches (Paul Churchland, Patricia Churchland, Edelman and others). The general idea is that the specific content of conscious experience (not only intentionality, but all "qualia") will be reduced by the progress of experimental science over old-fashioned notions of "folk-psychology", in exactly the same way as myths about astronomical phenomena have been erased by scientific astronomy[6]. We shall see later.

Returning to SAI, the assumption is that mental performances are equivalent to materially implemented computing operations and vice versa, without any "mentalistic" specificity. The first point, of course, consists in defining what a mental performance is. For SAI, this is generally outlined against a behavioristic background, its first explicit milestone being the philosophical experiment of the Turing Test. In this context, SAI's supporters (such as Daniel Dennett) say that only one argument could disprove SAI's claims: to indicate a mental act that (at least in principle) cannot be simulated by a computer.

---

[4] More generally, the identity theory assumes the identity of mental states and brain activities. We neglect here objections based on Leibniz' principle of the identity of indiscernibles.

[5] This entity is often ironically evoked as the typical non-scientific and false explanation of mental specificity.

[6] See Churchland P.M. 1996, Churchland P.M. and McCauley 1996.

Non-imitability, as the only possible proof of the failure of AI, is an idea that has often passed from supporters to opponents, giving rise to a large number of real and conceptual tests focused on creativity, the supposed distinctive core of "humanity". In my opinion, however, this concept of humanity as being committed to special performances too difficult to imitate is mistaken in principle, inasmuch as it accepts the behavioristic premise of coincidence between being and appearing, i.e. the very issue under discussion.

Curiously enough, this pragmatic attitude of computer-oriented behaviorism in assigning intentionality to artificial organisms coexists with the complementary attitude of pure behaviorists (Skinner, Quine) in denying any reality to *human* intentionality. Curious, but not illogical.

Section 5

To be more definite, consider a typical situation (discussed, for instance, by Daniel Dennett)[7]: the case of a chess player program on a PC. Omitting subtle nuances in Dennett's analysis, the crude conclusion is that all the features we require from a human player to say "he wants to win, he displays an intentional strategy, etc." can be repeated step by step in the case of the chess program, and that it *therefore* is an intentional system (where "it" means the diad: hardware and software, computer and program). We could render this example even more drastic by taking two (possibly different) computers endowed with (possibly different) deterministic chess programs: computer A's output is B's input, and vice versa. After the first move, everything is strictly determined. Can we still speak of intentionality in such a situation? The likely answer of SAI supporters would be yes: what you think is what you do (better: what It thinks is what It does). Moreover, due to strict determinism, such an intentional system would unmask the illusion of free will as an *external* cause.

First of all, we must observe that "determinism" in this context has nothing to do with the physical implementation of the game. The hardware side is obviously necessary but not sufficient to guarantee the existence of programs; their features (including determinism) do not emerge from below, from the hierarchical levels of lower complexity (the physical level / device

---

[7] See Dennett 1978.

level / assemblage level / low and intermediate languages levels, etc.). On the contrary, we have here a typical example of a determinism descending from above, from the programmer's world. Certainly, inasmuch as it is a dynamic system, a deterministic game may perfectly mimic the scheme of mechanical determinism (Rule plus Initial Conditions). But there is a deep difference: in the mechanical world of physics the rule is intrinsic, here it is extrinsic. The pre-quantum idea of the Universe as a global dynamical system, *à la* Laplace, with its totalistic[8] pervasiveness, is antithetical to the idea of a "machine". A machine, indeed, is not simply a mechanically operating reality, it is a device embedded in another and larger world, the world where it has been *intentionally* built. Analogously, the game does not exist *as a game* without reference to the players' world. The (intentional!) totalistic claim of reductionists needs concepts (machines, programs, games) which are intrinsically non-totalistic, or even anti-totalistic.

Second, if Dennett were right, the very identity assumption makes clear that there is no real relation or implication between intentionality and complexity, but only (and this is trivial) between complexity of programs and complexity of performances. Indeed, if the PC's intention of winning the game is one and the same thing as the chess program, then, for the same reasons, there is just as much intentionality in a kitchen appliance when it is whipping cream. The privilege assigned to playing chess with respect to whipping cream is simply due to an anthropomorphic, insubstantial aura (showing the pertinence, perhaps, of folk-psychology!).

Third, without the identity assumption (not conclusion), it would be quite natural to assign intentionality in the chess game not to the computers but to the observers: it is in their world (which is the same world as that of the programmers, and this is important) that we may read a strategy planning to win, because a strategy is such in the observers' interpretation, not in the material sequence of moves. As others have noted, in Dennett's and similar argumentation there is an implicit shift from the syntactical (operative) level of programs to the semantical (interpretative) level of observers.

The usual objection to the above observations is that no shift is necessary, because no interpretation is really required: for instance, following an example of Patricia Curchland[9], exactly as one says that

---

[8] By *totalistic* I mean a form of holism where everything, possibly including the theory itself, is in principle an object of the theory. In this sense, totalistic is not opposed to *reductionist* (on the contrary, scientific totalism implies a radical reductionism); it is opposed to *reduced.*

[9] See Churchland P.S. 1994 (quoted in Brigsjord 1994).

molecular motion *is* heat (and not *the cause of* heat), so one could say that the computer operations *are* their meaning. This is indeed a possible form of the identity theory. But such an analogy could work only if the computer operations had in themselves their sufficient reason (molecular motion has no extrinsic cause), while programs *have* an extrinsic origin. More: not only computers but every engine has extrinsic reasons for its operation. The jump ascending from syntax to semantics cannot be eliminated by a linguistic escamotage: external activation has been necessary to start these operations, which should otherwise be self-explaining, or self-interpreting. Physicalist comparisons are not correct for the same reasons which render a machine (every machine) distinguishable from purely physical or natural objects.

(I don't insist here on the fact that syntax itself, as Searle noted, cannot be reduced to pure physics).

Section 6

Two further and different objections can be expected.

The first: yes, apparently there is an extrinsic initiation, but this is only a technical deficiency of present days computers, and these difficulties shall be overcome in the future, when computers will be able to program themselves, making the identity scheme valid.

This objection may be answered by noting that self-programming computers (if there are any) will appear as the result of a previous chain of intentional operations that do not *emerge from* but are *immersed* or *embodied in* the computer world. If Artificial Intelligence is *artificial*, indeed, then the Artifex' imprinting remains as an ineffaceable seal.

In other words: if not a *res cogitans*, surely a computer is a *res cogitata*. Every machine is such. The strict comparison between artificial and biological brains could be reversed, leading in a direction opposite to the purposes of the SAI: not "biological brains are also a sort of computer", but "brains—or natural organisms seen as machines—are also *res cogitatae*". An interesting conclusion indeed, which stood at the basis of ancient natural philosophy; but not, I presume, the conclusion that a materialist would draw. Still, Michael Behe for instance has recently participated in a heated discussion in which he attempts to illustrate what he calls the "irreducible complexity" of living organisms at the biochemical level[10]. He argues that the elementary constituents of living matter *must* be seen as machines. It is precisely to prevent similar considerations that from the very beginning Darwinists (as natural philosophers, not as scientists) have been obsessed with giving relevance to the blindness of the selective pressure in its role as the *only* evolutionary force.

The misleading point is an ambiguous usage of "emergence", a term that suggests in advance a particular philosophical solution, which afterwards is mistaken for a "scientific result". If, from the very beginning, instead of emergence one had spoken of embodiment, then the computer parallelism would be a perfect representation of the opposite philosophical attitude: the independent existence of formal principles preceding their material realization[11]. I don't insist on this approach as "proving" anything; I only stress that *at least for computers*, whose coming into existence is explicitly known, the second attitude is by far more natural than the first.

Section 7

The second objection: yes, apparently there is an extrinsic start up, but since the human brain itself is a sort of biological computer, all the identity arguments are still valid, provided that the programmer (as a materially implemented function, not as a person!) is included to complete the computer-system.

This objection may be answered by noting the circularity of the reasoning: the validity of the identity theory is proved by assuming its validity. Differently stated: SAI's claims are proved by using SAI's claims.

However, this second objection deserves further comment. If not a proof, indeed, it could be a good *illustration* of the SAI's point of view, and in this respect we can examine its self-consistency. Instead of the usual diad (hardware-software, or machine-program) we now have to do with a triad: the whole computer-program-programmer system. Let's admit that this *is* an intentional system due to the complexity of the biological component, the human brain seen as a computer. In this picture, intentionality is a sort of

---

[10] See Behe 1996.

[11] An attitude we could consider Platonist, in a broad sense. It is compatible with functionalism.

self-referential function of the system, inasmuch as the system must be intentional in order to recognize (or to assign) intentionality. Is it possible to speak properly of "emergence" in such conditions? Then, when a brain recognizes itself as an intentional system, this means that in it there exists a further (and sufficiently complex) intentional shell that recognizes the rest of the brain as an intentional system. Now, either this is only a metaphorical way of speaking, or such a shift to an upper interpretative shell (the shell of the subjective "who") in the brain-computer means that the semantical level cannot be objectified. Objectification would indeed require another shift to a further shell, and so on: such an operation implies, logically, a *regressus ad infinitum*, or, materially, an explosive growth of complexity.

Section 8

Previous discussion may also be useful in seeing the fallacy in those hopes of obtaining a definitive understanding of the mind-body problem based on the progress of the neurosciences, along the lines espoused by, for instance, eliminative materialists. We must of course distinguish these philosophical claims from the concrete work of neuroscientists. On a philosophical level, complexity often plays more of a rhetorical than functional role: it guarantees that there are many steps before the goal of a complete understanding is reached, leaving space for the hope-based eliminative ideology. Now, the fallacy of this ideology does not consist in the possible limits to the scientific knowledge of the brain. Patricia Churchland is certainly right in asserting that the argument of present ignorance cannot be taken as an indication of future impossibility[12]. But the computer parallelism makes clear that the question is flawed in its very origin. For machines, indeed, the "neurosciences" coincide with computer engineering, therefore *we already have* a complete and perfect knowledge in every detail of the working principles of a computer: we will hardly in the future obtain a comparable knowledge of the human brain. Notwithstanding this, as the discussion of the intentionality of chess players has shown, the solution is entirely committed to philosophical attitudes. It is not true that intentionality or other mental items (e.g. the nature and ontological constitution of qualia) will be erased by progress in the neurosciences. In the same way, I don't

---

[12] See Churchland P.S. 1998.

expect that a complete understanding of the working principles of a car may reveal why this car goes south or west, left or right (even if every choice in going south or west, left or right, has a physical and understandable realization).

Section 9

On grounds even more empirical, the neuroscience approach, independently of the enormous scientific and practical interest of this research, is subject to the same criticism made of SAI: a mapping from mental functions to the activity of the brain does not define the "who", which is arbitrarily assigned by external pragmatic exigencies. Neurological achievements and improvements could fit an "embodiment" paradigm *at least* as well as emergentism. Therefore they do not prove anything with regard to our problem. Moreover, the flexibility of the brain in performing the same mental tasks with different components (as observed, for instance, in cases of surgical mutilations) seems to exclude a rigid correspondence component-function (providing, perhaps, some arguments for functionalists...).

It is noteworthy that complexity, in these studies, is sometimes introduced *ad hoc* to make plausible the emergentist claims. The strategy is as follows:

a) Choose an object (brain, computer, programs, ...) putatively responsible for mental performances.

b) Define "complexity" in such a way that the resulting object is sufficiently and effectively complex.

c) Assign this complexity as the material cause of mental performances.

Now, this causal dependence is arbitrary, since it is completely dependent on the initial decision.

Edelman and Tononi, for instance, substantially adopt this strategy[13]. The complexity they introduce is an interesting quantity, based on experience doubled with numerical simulations. Notwithstanding some technical difficulties[14], we can accept their definition as reasonable. However, this apparently empirical conclusion simply confirms an initial

---

[13] See Edelman and Tononi 2000.

[14] The operative definition of probability measures, necessary to introduce entropy, cannot be extended in an obvious way from finitist models to the realistic case.

persuasion: the brain is a very complex apparatus (and we all agree!). But the causal link between consciousness and complexity remains totally conjectural since, once again, each of their empirical statements would be equally true within an embodiment paradigm.

For clearness: Edelman and Tononi criteria for complexity (differentiation in functions, high specialization in performance, long range correlations, etc.) could easily be applied to an orchestra, but the conclusion that this complexity is a sufficient reason for music would be completely arbitrary; and it would remain arbitrary even for an increasingly complex orchestra with billions of musicians.

Section 10

We must resume the observation of section 3: to speak about complexity requires a preliminary symbolization of the objects to be investigated in a formalized context, revealing the observer's interest. This is not at all a new situation in science (something very similar happened, in particular, with "information"). We know that science does not apply to the singularity of empirical facts, but to idealized objects defined in a (more or less explicit) theoretical context[15]. An apple is a scientific object inasmuch as it is seen, for instance, as a body with mass, not as the only apple in the history of Universe with *that* shape, *those* brown spots, *that* worm[16]. Now, conscious experience, including intentionality, is singular in its essence. Certainly, out of consciousness symbolic representations of formalized and idealized classes of experiences may be drawn: but such representations belong *eo ipso* to another order of reality. All efforts to investigate consciousness are not about "the real thing" but about these symbolic representations of abstract, idealized classes of thoughts. The problem has been shifted; the original object destroyed. Perhaps this is the only practical approach to the matter, but it is not a proof of the reducibility of consciousness.

---

[15] The idea of repeatability rests on this.

[16] And this does not imply that the shape, spots and worm are unreal!

Section 11

The above misconception coheres to a frequent, spontaneous identification of "consciousness" and "self-consciousness". Self-consciousness, at least in modern western psychology, is the same as self-representation. This means that the apex of consciousness is frequently (wrongly) identified with an objective state of affairs, already caught in the frame of linguistic formalization (with the further paradoxical consequence that the Subject of self-consciousness would be unconscious!). Once this is accepted, it is not surprising that the "scientific approach" to consciousness and its phenomenology treats of something which is essentially different from the original fact one pretended to explain. Usual objections, such as "pre-linguistic conscious experience is only subjective", are misleading. They confuse subjectivity of judgment ("I like ice cream") with subjectivity as to function: the "real thing" to be explained through being made conatural with intentionality. Moreover, these objections are self eliminating: since every representation is a second-degree reality with respect to the represented entity (at least in the materialistic frame we are exploring), the non-reality of conscious experience would imply the non-reality of its representation (surviving, at most, as a cultural convention in a sociological sense, not as a scientific object in the physicalist sense).

In fact, the impasse is usually overcome by stressing the social value of representations: social exchange, technological simulations, and all that; but this is not in the realm of scientific explanation.

Section 12

A way out of the flaws of the SAI and neuroscience approaches chosen by John Searle involves biological naturalism[17]. In a famous comparison I have already cited, consciousness would emerge from the brain as liquidity from a large amount of H_2O molecules: a collective property not had by single constituents.

First of all, we observe that in this case, finally, we may properly speak of emergence. Second, what Searle discusses is "the real thing": consciousness as it is experienced, not consciousness as it is represented. On

---

[17] See Searle 1992.

the basis of this correct starting point, Searle can develop a reliable criticism of other approaches (SAI and eliminative materialism in particular). Unfortunately, this solution is only apparent. We indeed have to do with a dilemma: either Searle is tautologically asserting that complexity is the need to have what you have; or, if his complexity is explanatory, that we should expect the emergence of other intentionalities in his sense from all sufficiently complex apparata. A large town, for instance, or the Amazonian ecosystem, are certainly complex systems. Are we ready to ascribe a proper intentionality to a forest or to a town? (I stress: to a town, not to the town's mayor!) Following this line, one would finally meet such things as Hobbes' Leviathan, or the "Gaia hypothesis" (the materialistic version of the old traditional concept of *Anima Mundi*), or other exotic psychisms. It is an instructive paradox that a sober and rational approach such as Searle's finally proves to be either empty of real scientific content, or unwillingly close to ambiguous positions shared by New Age supporters!

Perhaps this is a proof that consciousness is a dangerous matter.

References

Behe, M.: 1996, *Darwin's Black Box*. The Free Press.
Brigsjord, S.: 1994, "Searle on the Brink", *Psyche* 1, 5.
Churchland, P. M. and McCauley, R. N.: 1996, *The Churchlands and Their Critics*. Blackwell Publishers.
Churchland, P. M.: 1996, *The Engine of Reason, the Seat of The Soul*. The MIT Press Books.
Churchland, P. S.: 1994, "Can neurobiology teach us anything about consciousness?", *Proc. APA* 67, 4.
— 1998, "Brainshy: Nonneural Theories of Conscious Experience", in S. Haneroff, A. Kaszniak, A. Scott (eds.): 1998, *Toward a Science of Consciousness II*. Cambridge MA: MIT Press.
Dennett, D.: 1978, "Brainstorms". Bradford Books.
Edelman, G. and Tononi, G.: 2000, *A Universe of Conscience*. Basic Books.
Grush, R. and Churchland, P. S.: 1995, "Gap's in Penrose's Toilings", *Journal of Consciousness Studies* 2, 1.
Li, M. and Vitànyi, P.: 1997, *An Introduction to Kolmogorov Complexity and Applications*. New York: Springer.
McCarthy, J.: 1983, "The Little Thoughts of Thinking Machines", *Psychology Today*.
Penrose, R. and Hameroff, S.: 1995, "What 'Gaps'?", *Journal of Consciousness Studies* 2, 2.

Penrose, R.: 1994. *Shadows of Mind*, Oxford University Press.
Searle, J.: 1992, *The Rediscovery of the Mind*. Cambridge MA: MIT Press.

12. Can Supervenience Save the Mental?

Luisa Montecucco

Dept. of Philosophy, University of Genoa, Italy

Our everyday understanding of each other ₃ ₔₔ discourse" which is based on intuitions associated with daily experience, on chains of thoughts and reasons, on meta-reflections about feelings and thoughts and reasons. Such a type of discourse takes the standard shape of "propositional attitudes" ascriptions: human beings explain intelligent actions pairing mental attitudes, like believing, hoping, fearing etc. with specific propositions (*e.g.* "it will rain"). I may explain my (or somebody else's) reluctance to water the garden by mentioning the belief or the hope that it will rain, or the fear that it will rain "cats and dogs". Common-sense psychology works well in explaining and predicting human behaviour, as if it were (or since it is) a causally potent theory, which posits unobservable things like beliefs and intentions and connects them through law-like relations. What are then propositional attitudes, how can they cause human beings to act in appropriate ways, what could make that theory true (if it is)?

The tentative answers of contemporary cognitive science mainly tell "(...) a fully materialistic story in which mindware emerges as *nothing but* the playing out of ordinary physical states and processes in the familiar physical world"[1]. What is special about the mental, from the nature of mental states to the power of thoughts to cause further thoughts and actions, receives a broad range of much debated explanations concerning the structural properties of the system in which thoughts occur. To support the

---

[1] Clark 2001, p. 3.

thesis that our reason-respecting mental life is just an appropriate organisation of matter, a variety of sciences of the mind with different models and methods are cooperating, in a state of very rapid and continuous development and sophistication. If the more than half a century old notion of the brain as a kind of computer—based on some equivalence between intelligent behaviour and information processing[2]—now seems too simple, work on connectionism and neural networks, artificial life (*Alife*), situated cognition, robotics faces the once neglected neural and ecological aspects of the "thought organ" as a radically different sort of device, not necessarily— or not only—computational.

However, the conceptual and technical transition from the logical scheme of a Turing machine as a model of intelligent behaviour to biologically oriented artificial neural networks—and other kinds of models working and "learning" at a sub-symbolic level[3]—does not seem to solve or even challenge all the problems of the specificity of the mental with respect to the physical. They are related both to the heritage of a millenium-long tradition of philosophical inquiry that apparently resists a process of "naturalisation,"[4] and to everybody's present consciousness of their own mind and its related common-sense notions, which together seem to resist a process of "deleting." Of course it is very difficult, if not impossible and even conceptually incorrect, to advance generalizations over such a heterogeneous profusion of ongoing research programmes: what can be safely said is perhaps that the supposed "orthodoxy", represented by varieties of materialistic monism, needs to be, and in fact is, corrected and rearranged to integrate possible solutions to the problems regarding the mental, about which we shall speak.

---

[2] In 1948 Alan Turing was already speaking of "thinking" and "intelligent" machines and this vision inspired classical Artificial Intelligence (AI).

[3] For an interdisciplinary view on the relevant scientific research during the second half of the twentieth century, Cordeschi 2002 presents historical and theoretical issues of cognitive science, AI, psychology, neuroscience and the philosophy of mind. A collection of basic papers illustrating the transition from classical AI paradigm to connessionism is Haugeland 1997.

[4] As an introduction to contemporary naturalism, which also debates whether a naturalistic perspective can accommodate apparently non-natural features such as normativity, see Villanueva (1993).

2. Emergence and evolution[5]

Classical cognitive science, on the basis of the analogy between the mind and the computer software, had made use of different explanatory categories to interpret mental and physical phenomena, trying to avoid both substance dualism and mind-brain identity. In this perspective, to explain the nature and the contents of mental life what matters is the formal organisation of the system linking possible inputs, inner computational states and outputs (actions, speech), not the potentially variable material support, such as the human brain and the computer hardware. Psychological states (or at least some of them) are then considered to be *multiply realisable*, i.e. the same psychological state can be realised by different physical supports (or "hardware" organisations).

The residual dualistic stance of *functionalism*[6] is at least possibly overcome by an orientation often defined as "emergentism". In general, many systems have behaviour (*emergent* properties) which result from the collective behaviour of the system's components rather than from the action of any single component: these properties are *unanticipated* and, in the case of artificial systems like the ones studied in Alife, unplanned or unprogrammed[7]. In nature, abound examples of emergent properties, but perhaps nothing resembles the truly amazing leap from the highly complex neuro-physiological organisation of the human body and brain and the manifestation of cognitive capabilities and intelligent behaviour. Cognitive

---

[5] Already in 1923, Lloyd Morgan—one of the British emergentists, active in the first half of the last century—connected the two notions in his *Emergent Evolution*, claiming that through the process of evolution new properties emerge in ways unanticipated by the laws governing matter at lower levels of complexity. British emergentism developed the distinction, drawn by John Stuart Mill in his *System of Logic* (1843), between the "mechanical mode" in which two or more causes combine to produce a ("resultant") effect which is the sum of the effects of each cause acting alone and the "chemical mode" in which this does not happen, producing effects later called "emergent" by George Henry Lewes (1875). Mill spoke of "homopathic effect" in the first case and of "heteropathic effect" in the second one.

[6] See Putnam 1967, one of the manifestos of functionalism, containing the thesis of the "multiple realisability" of mental states.

[7] The use of biological ideas in computer science, initiated by pioneers such as John von Neumann, Alan Turing and Norbert Wiener, continues today in a field called "evolutionary computation" (EC), whose methods are loosely inspired by biological evolution. EC methods—for instance, "genetic algorithms" (Holland 1975)—have been used in models of natural systems in which evolutionary processes take place, as in the case of the interaction between evolution and cognitive processes. Many of these models constitute the research field of Alife.

scientists and philosophers frequently share the idea that physical and mental properties do not belong to different ontological spaces, but the latter ones emerge as an effect of the "complexity" of the former ones.

Let us recall a few very general qualifying aspects of *complex systems*: high sensibility to initial conditions; several interacting elements; non-linearity of interactions (missing proportionality between causes and effects); emergence, at the global level, of properties not detected on the local one; transition to different qualitative states as a consequence of the amplification, even casual, of small perturbations; self-organisation (emerging of structures endowed with regularities and symmetries without the involvement of causes external to the system). As Agazzi notes in his contribution to the present volume, complexity is not an intrinsic property of a system, but depends on the perspective adopted. Science does not renounce the description of complex phenomena in terms of regularities and ordered structures as well. Self-organisation and state re-configurations in crucial moments of their evolution clearly show in these systems the presence of an underlying order, even if of a dynamic nature. The peculiarity of non-linear phenomena necessitates a revision of models of causality but, at the same time, does not preclude a reductionist approach[8]. Of course, the observation of phenomena evolving in an unexpected *creative* way opens the problem of finding laws capable of explaining emerging regularities and of answering the basic question: which are the critical mechanisms generating *that* specific order? For instance, at the basis of complexity, as well of computability, there seems to be the mechanism of recursivity: from recursive circularity of cause-effect relationships often emerges self-organisation, even in absence of something like a "project". However, recursivity by itself is a purely mechanical process and the expression "self-organisation" can lead to misunderstanding and ambiguity, due to its latent anthropomorphic interpretation. On the other hand, since there is a natural recursive mechanism (the duplication of DNA in the cell) at the very basis of life, both scientists and philosophers advance intriguing questions such as: are there limits to levels of complexity reachable through a recursive

---

[8] Actually, the British emergentists intended emergence as implying irreducibility. However, the Newtonian conception of mechanistic reduction they endorsed was soon broadened by quantum mechanics: the quantum mechanics explanation of chemical bonding—a truly reductive explanation—led to the declining of antireductionist, emergentist view of chemistry and biology (McLaughlin 1992).

(algorithmic) apparatus? Are there aspects of intelligent behaviour that can be recursively generated?

The "new" cognitive science, with the increased biological plausibility of second-wave connectionism[9], attempts to explore possible answers in ways more cooperative towards sciences concerning human beings and their behaviour, in particular the biological sciences. For instance, within this horizon, *evolutionary psychology* theorizes that the complex cerebral structure from which our psychological faculties emerge has undergone—like any other physical organ—a phylogenetic evolution as a consequence of its interaction with the environment. The adaptive process triggered by natural selection is supposed to have modified both the neuro-physiological structures and the cognitive structures emerging from them: the mental and the physical would evolve together interacting and shaping each other, being two aspects of the same material reality. However, the theoretical apparatus of evolutionary biology doesn't seem sufficient to allow a real understanding of how the mental and the physical interact. Neurobiology and genetics will have more to say about the emergence of the mind—at the ontogenetic and at the phylogenetic level—from the neuro-physiological properties of the brain, in correlation with the genetic information stored in the DNA. What is unusual and noteworthy is that evolutionary psychology aims at constructing a comprehensive map of the species-typical computational architecture of humans, including not only cognition but also motivational and emotional mechanisms.

It should be quite evident by now that two aspects of the notion of emergence interplay and need to be distinguished, as Ernest Nagel did in the classic *The Structure of Science* (1961), since the novelty of emergents can be construed with respect to time—in historical, evolutionary, cosmological processes—or with respect to ontology[10]. In the first case, emergents are the first occurrences of whatever the emergent is supposed to be (*diachronic emergence*); the variety of the universe is the result of a "creative" development from a primitive stage towards unforeseen novelties. In the second case, emergents are something new coming to existence with each

---

[9] Connectionism, parallel-distributed processing, artificial neural networks identify partially overlapping research programmes that bear some relation to the architecture and workings of the biological brain, unlike classical AI view of the mind based on symbolic computation, serial processing and a strict relation to language and logic.

[10] To be precise, Nagel regards the doctrine of emergence as essentially correct if interpreted as a thesis about *logical* relations between propositions. However, it is the ontological point of view that has prevailed in the subsequent literature on the topic.

instance of a pattern of lower level constituents (*synchronic emergence*), according to a non-reducible hierarchical organisation. Even if the two aspects are closely related (temporal emergents would be the first instances of particular ontological emergents), diachronic emergence can be treated as an historical problem, while the stronger notion of synchronic emergence poses logical and ontological problems. It will be mainly this second notion that we are going to associate with the one of "supervenience", without following the development of a "neo-emergentism" in the writings of such well known philosophers as Joseph Margolis and Edgar Morin (in the so called "continental" tradition) or Karl R. Popper and, recently, Douglas R. Hofstadter (in the "analytic" tradition, but with opposed ontological perspective, being Popper close to dualism and Hofstadter explicitly materialist).

3. From emergence to supervenience

The notion of an emergent property of a whole as a non-additive resultant of properties of the parts of the whole—a notion developed with reference to chemical, biological and finally psychological phenomena—has a quite strict relation with the philosophical notion of *supervenience*, introduced into the philosophy of mind by Donald Davidson. Explicitly following moral philosophers G.E. Moore and R.M. Hare, he generalised the idea that there could be no difference in moral respect without a difference in some descriptive, or non-moral, respect:

> ....mental characteristics are in some sense dependent, or supervenient, on physical characteristics. Such supervenience might be taken to mean that there cannot be two events alike in all physical respects but differing in some mental respect, or that an object cannot alter in some mental respect without altering in some physical respect. Dependence or supervenience of this kind does not entail reducibility through law or definition...[11]

Does this mean that mental entities are identical with physical entities? and that there are rigorous psycho-physical laws? Davidson answers "yes" to

---

[11] Davidson 1970,1992, p. 141.

the first question and "no" to the second one: according to his *"anomalous monism"*, there is only one class of entities, but the possibility of definitional and nomological reduction is denied. He contrasts his view with "nomological monism" or materialism (the two types of entities are identical and correlated by strict laws), with "nomological dualism" (in the traditional variants of interactionism, parallelism, epiphenomenalism) and with "anomalous dualism" (there are no laws correlating the mental and the physical, regarded as different substances). *"Supervenience" is then related to monism, but without reductionism*, at least as it is intended by its proponents, engaged in saving what seems to be the unavoidable physicalistic output of contemporary scientific advancement and, at the same time, what folk (common-sense) psychology is not willing to give up, *i.e.* the specificity of mental reality. To reach this dual objective, Davidson claims that *the mental is not an ontological but a conceptual category*. In his view, mentalistic discourse about thoughts and intentional actions does not imply the necessity of giving an autonomous ontological status to special entities of a non-physical nature and present in the "mind". As a matter of fact, mental objects and events are at the same time physical, chemical, biological and psychological objects and events. What changes is the vocabulary in which we describe them: for instance, the mark of mentalistic vocabulary is semantic intentionality.. Reason-explanations differ from physical explanations because they contain an *intentional vocabulary*, the basic concepts of which cannot be reduced to the vocabularies of the physical sciences, primarily because of their *normative* character. It follows that psychology cannot be reduced to physics, nor to any other of the natural sciences. In more recent Davidson words, then,

> ....the relation between the mental and the physical, which Quine now seems to accept, is what I have called anomalous monism, the position that says there are no strictly law-like correlations between phenomena classified as mental and phenomena classified as physical, though mental entities are identical, taken one at the time, with physical entities. In other words, there is a single ontology, but more than one way of describing and explaining the items in the ontology...[12]

---

[12] Davidson 1995, p. 4.

The single mental event, then, corresponds to some physical event (*token identity* theory). But this does not mean that another occurrence (token) of that mental event should correspond to the *same* physical event (this would be the stronger *type identity* theory). With this, however, Davidson doesn't deny that there are regularities or type/type law-like generalizations, even if such regularities cannot be transformed into "strict"[13] laws connecting events described in mental terms with events described in physical terms. The problem is that only physics aims at providing a closed system governed by strict laws, while—in the case of psychology—mental types (and the fact that an event token belongs to one of them) depend on certain background assumptions about meaning and rationality, placing a normative constraint on the correct attribution of propositional attitudes.

It is difficult to evaluate this intriguing argument, particularly within the limits of the present inquiry, which do not allow space for the consideration of some widely discussed criticisms of anomalous monism, for instance its failure to explain mental causation, since only events under a physical description are causally effective.

Given the conceptual framework of this volume, let us instead return to the concept of supervenience and to its allegedly monistic ontological foundation. *Is the "anomaly" of Davidson monism sufficient to allow us to take mentalistic discourse seriously* in such a way that—within that explanatory apparatus—what we call "mind" is not deprived of its essential faculties and properties, the object of human beings unvarying intuitions[14]? *And, if it is sufficient, should we still call it ontological monism?*

In trying to answer these two basic questions, we make the choice of proceeding to follow the development of the notion of supervenience that has then played a key role in the philosophy of mind during the last two or three decades, especially thanks to the work of the philosopher Jaegwon

---

[13] A "strict" law is an exceptionless general truth that makes no use of open-ended escape clauses such as *"ceteris paribus"*—other things being equal. Strict laws must belong to a closed system: whatever affect the system must be included in it. The problem with *ceteris paribus* laws is that their testability is reduced as the number of possible interfering factors to be held constant grows; this makes too easy for anyone to claim to have uncovered a scientific law. How is one to determine whether other things are indeed equal, when there is an inexhaustible list of them? Astrology is a good example for understanding the immunity from disconfirmation of laws employing *ceteris paribus* clauses.

[14] The risk laying just a few conceptual steps from anomalous monism is plain "reductionism" and even "eliminativism" or "eliminative materialism", the view that our common-sense conception of the mind is basically mistaken and should be dropped in favour of a scientific one which does not refer to propositional attitudes such as beliefs and the like.

Kim. Three features seem to characterise the relation between a set of supervenient properties and a set of "subvenient" or "base" properties:

1) if two things cannot be distinguished in terms of subvenient—in our case, physical—properties, they must be indistinguishable also in terms of supervenient—mental—properties (*property covariation*);

2) supervenient properties are dependent on—determined by—related subvenient properties (*dependence*);

3) supervenient properties are not reducible to their base properties (*non-reducibility*), even if this is of course not necessarily implied by property covariation and dependence.

The British emergentist already mentioned in note n. 5, Lloyd Morgan, used "supervenient" as a linguistic variant of "emergent", with significant similarities showing the acceptance of the three features, on the basis of an initial systematic formulation of *non-reductive physicalism*: in fact, emergent properties were considered by Morgan to be "genuinely novel" with respect to the lower level properties which they depend on and emerge from, when a certain degree of complexity of an appropriate kind is achieved.

However it has been pointed out that the focus of the theory of supervenience and the focus of the theory of emergence are to a certain extent different[15]. In fact, the former places the emphasis on the notion of *dependence*, since it aims at finding whether there is a dependency relation weak enough to save the non-reducibility of the mental to the physical and hence compatible with the non-existence of psycho-physical laws. As far as the theorists of emergence—such as C. D. Broad[16]—are concerned, the notion of *reduction* was of outmost importance: for them the dependence of the mental on the physical can only consist in the existence of laws connecting them. Otherwise it would be impossible to give an account of emergent mental properties by appealing to their corresponding physical properties, and thereby to found the possibility of mental causation. At the same time the pure existence of bridge laws is neither a sufficient nor a necessary condition for reducibility: in fact they could attest *whether* there are lawful connections between the mental and the physical, but not explain *why* they exist, if they do. *Microreduction*, on the other hand, shows why objects with a certain microstructure *must* have certain macroproperties, but this does not preclude the possibility of emergent properties, since their decisive feature is that they "...cannot, even in theory, be deduced from the

---

[15] Beckermann 1992.

[16] Broad 1925.

most complete knowledge of the properties of the components $C_1,..., C_n$ in isolation or in systems with a different microstructure"[17].

4. Strengths of supervenience

Having drawn attention to some interconnected conceptual aspects of the notions of emergence and of supervenience—also through a few historical references and an explicit link to the context of contemporary philosophy and cognitive sciences—the time has come to try to answer to the question that constitutes the title of this paper. Actually, so far all our steps forward have raised questions rather than answer those already before. It is not even clear whether we are following an inquiry on the logico-epistemological or the ontological level: for instance, Broad's words, reported at the end of last section, look upon non-deducibility as characterising emergent properties in such a way that emergence seems to work on the logico-epistemological level and to concern our knowledge both of components and of wholes. Whatever Broad's intentions, the suspicion arises that emergence be a way of expressing our lack of understanding, something like *a measure of our ignorance*. Why should it be that what is unforeseen and novel for a human observer corresponds to something novel on the ontological level? Or are the whole significance and purpose of philosophical emergentism to escape the boundaries of a totally material world through the appearance of novel emergent properties, which can at least in a sense be distinguished from physical properties in accord with strong common-sense intuitions?

From the previous questions, others arise at a deeper level, questions which cannot receive answers here. In referring to different conceptual context and different authors, we used with a certain freedom—relying on shared philosophical terminology or even on commonsensical discourse— linguistic expressions such as: substances and their properties, entities, facts, events and states, etc., to oppose the mental and the physical. What is the metaphysical nature of events? Are substances ontologically prior to events? Answers to these and similar questions, underlying all research on mentality, of course seriously affect a theory of supervenience.

Let us now follow Kim's inquiry in some detail, an inquiry he has pursued over a period of approximately thirty years since the early 1970s.

---

[17] Broad 1925, p. 61.

The core idea of mind-body supervenience, as expressed by Davidson, is that *indiscernibility with respect to physical properties entails indiscernibility with respect to mental properties..* It follows that an exact physical duplicate of myself would necessarily be a psychological duplicate as well. Kim[18] has shown that the relation of supervenience can be explicated in various not equivalent ways and be characterised by varying "strengths". In order to review his definitions, let us assume the existence of a not empty domain D of individuals (persons and other psychologically interesting organisms or structures) and of two sets of properties defined over D: M is a set of mental properties and P a set of physical (biological and physico-chemical) properties. A "state" can be thought of as an individual's instantiating one or more properties at a time; an "event" is the changing of a state with respect to these properties. Given that supervenience of events and states can then be equated with property supervenience, *what is it for M to supervene on P over D?* The answer depends, among other things, on whether we only compare things that come from one "possible world",[19] or whether we also compare things that come from different worlds. If we consider a plurality of worlds, we could prohibit there being in each of them physical duplicates that are not mental duplicates. This corresponds to what Kim calls *"weak supervenience"*, defined thus:

Necessarily (i.e. in every possible world), if any x and y in D are indiscernible[20] with respect to P, then they are indiscernible with respect to M.

The relation is meaningful and not trivial if the set P is appropriately circumscribed, otherwise no two individuals in D would be alike with

---

[18] Among Kim's writings mentioned in the Bibliography, Kim 1990 contains an historical survey of the concept of supervenience, together with an overview of the philosophical debate over its different characterisations.

[19] The theory of possible worlds provides truth conditions for modal claims of possibility and necessity, which are integral to the notion of supervenience. From the ontological point of view, at one extreme is the position that possible worlds are just as real as the actual world and at the other extreme the position that possible worlds are no more than linguistic descriptions of how the world might be.

[20] Also the notion of *indiscernibility* is higly problematical: unless an empirical procedure for identifying two individuals (objects, states, events...) is available, they must be considered *epistemologically different*. If there are criteria for differentiating their intrinsic nature and their relations, they must be considered *ontologically different*. In order to legitimately identify them, there must be explicit methods and criteria allowing to claim their identity.

respect to P (for instance, they would differ with regard to their spatiotemporal locations). The *intra-world* constraint of weak supervenience does not by itself make impossible the existence of worlds exactly like the actual world in all physical respects, but in which a) mental properties do not emerge; b) organisms other than humans are conscious; c) everything is conscious. This unavoidable corollary of weak supervenience means that it is not strong enough to support the physicalist requirement of the *dependence* of the mental on the physical in such a way that the latter determines all the aspects of the actual and any possible world; it seems to agree instead with the *multiple realisability* of psychological characteristics.

Generally speaking, this should be

> ....a presumptive desideratum on the explication of supervenience: base properties must *determine* supervenient properties in the sense that once the former are fixed for an object, there is no freedom to vary the latter for that object.[21]

To achieve this stronger constraint, indiscernibility could be evaluated with regard to worlds taken as units rather than to individuals within worlds. This is the thesis of "*global supervenience*":

> *Any two worlds indiscernible with respect to P are indiscernible also with respect to M (indiscernibility of worlds with respect to a given set of properties means that properties are distributed over their individuals in the same way).*

Under this interpretation, worlds not distinguishable from the physical point of view cannot be distinguished from the mental point of view (of course the opposite is not true). Again we have multiple realisability[22] with a significant dependence of the mental character of a world on its physical character. Another step towards a definition implying fewer degrees of

---

[21] Kim 1990,1993, p. 60.

[22] Since the idea of multiple realisability is strictly connected to the functionalist conception of the mind, most functionalists are committed to some form of mind-body supervenience.

freedom (as far as dependence is concerned) takes us to *"strong supervenience"*[23]:

> *For any two individuals x in world w' and y in w", if x is indiscernible from y with respect to P, then x is also indiscernible from y with respect to M.*

Strong supervenience differs then from weak supervenience since it compares individuals not only intra-world, but also *cross-world*; and it differs from global supervenience since it applies indiscernibility considerations to individuals, *locally*, rather than to worlds, globally. Strong supervenience quite clearly implies weak and global supervenience.

Each one of the three previous definitions of supervenience—as well as other—

> may have its own sphere of application, serving as a useful tool for formulating and evaluating philosophical doctrines of interest. And this does not mean that we must discard the core idea of supervenience captured by the maxim "No difference of one kind without a difference of another kind"[24]..

This kind of dependence, however, does not entail reduction, or at least it is not supposed to, in agreement with the intention of the physicalists who oppose *psychophysical reductionism* (type physicalism). This last position requires, in very general terms, that for each mental property M there be a physical property P such that necessarily M is instantiated by an individual in D at time t *if and only* if P is instantiated by it at t. The reduction can be taken as *definitional* (as for logical behaviourism)—if the goal is to provide each mental predicate with an analytically equivalent definition in terms of physical-behavioural expressions—or as *nomological*, if it is carried out inter-theoretically in empirical science (an example is the reduction of optics to electromagnetic theory). According to this last model, mind-body reduction would then imply the derivation of psychological laws from physical (neurobiological) laws, through some empirical "bridge laws"

---

[23] We do not consider the alternative version of strong supervenience in terms of the modal operator of necessity, since it is provably equivalent, under certain assumptions, to the possible world formulation just reported (Kim 1987,1993).

[24] Kim 1990,1993, p. 155.

correlating mental kinds with physical-neural kinds. *This is exactly what (global or strong) supervenience is devised to rule out*, with a much debated potential for success.

Of course, a philosophical perspective can only consider and eventually show the logical and metaphysical (im)possibility of property- or theory-reduction, while it is up to science to generate actual reductions: the very idea of psychological properties supervenient on physical properties should be a contingent matter, to be debated on an empirical basis (still a long way to go). The theoretical point at which we have just arrived is rather that—to accommodate in a theory of supervenience a dependence relation robust enough to satisfy physicalism (moving from weak to global to strong supervenience)—the hypothesized connection of the mental with the physical is dangerously close to the reductionist connection. Under strong supervenience, physical duplicates are also psychological duplicates: they are duplicates *tout-court*. Considering the very incomplete and approximate scientific knowledge of specific correlations between neural and mental properties, the theory of mind-body supervenience seems (or *needs*) to rely on deep metaphysical commitments regarding the primacy of the physical. At the same time, it does not say anything about *what kind* of dependence relation is involved and therefore it cannot be considered a proper mind-body theory, as Kim explicitly acknowledges. According to Kim, a promising explication of the notion would be in terms of a specific kind of *"mereological supervenience"*, *i. e.* in terms of the dependence of the (macro) properties of a whole on the (micro) properties of its components— back to the key idea characterising emergence. Psychological properties would then be explained as macro-properties of a whole organism appropriately co-varying with micro-properties regarding the organism's constituent systems, organs, cells etc. Can we hope to find and explain the mental as emerging from the complex system of neurobiological phenomena? This is, very synthetically, how Kim could answer:

> Whether such microstructural explanations really "explain" mentality in the sense of making mentality, in particular consciousness, intelligible—something that the emergentists despaired of ever attaining—may be another question. Still, it may well be that mentality is best thought of as a special case of mereological dependence and determination[25].

---

[25] Kim 1993, p.168.

5. Can supervenience save itself?

The main problem of this short exploration, now approaching its end, has been the one of following a path, inside the very rich and fluctuating conceptual landscape surrounding the mind-body problem, that be narrow[26] enough to allow a brief specification of a question and at the same time meaningful enough to afford it a possible answer. The notion of supervenience, primarily considered from the point of view of the dependence relation between subvenient and supervenient properties, has supplied the guideline. Quite a few challenging conceptual knots have already been pointed out, without really crossing the threshold of a real theory of mind of a commonsensical or philosophical or psychological nature. In the previous sections, we highlighted difficulties related to the intended key role of supervenience in accounting for the dependence of the mental on the physical, without a reduction of one to the other, in a monistic ontological perspective. Doubts have been raised about the feasibility of saving non-reducibility and monism at the same time.

Moreover, the rather naive and implicit assumption of a kind of "substance metaphysics" to assert materialism has been recognized not only as a theoretical obstacle in defining supervenience, but also as having decisive effects on the integration of life and mind with the rest of the natural world. If all phenomena can be understood as the result of processes involving atoms and molecules—if "microreduction" is then a practicable option—mind could be nothing other than an epiphenomenal manifestation of fundamental-particle interactions. Mental phenomena would then be mere appearance, without any causal effects, since only physical phenomena are causally efficacious. It is here that the doctrine of supervenience should show its "strength", vindicating common-sense intuitions about the fact that mental phenomena exist insofar as they *make a difference*, having causal power and interacting with physical phenomena.

Let us suppose that mind-body supervenience is a basically correct theory: is there empirical proof that it holds for the stunning variety of mental states and events? Of course not. (What do we know about the exact neurochemical state subvenient on mixed feelings of relief and satisfaction,

---

[26] As a matter of fact, it has been necessary to adopt a few oversimplifications and to leave out important references in the philosophy of mind (for instance, to the well known contributions of Saul Kripke and of John H. Searle, or, more recently, of David J. Chalmers, among many others). However, the overview of the problematic dependence of the mental on the physical, as stated by the relation of supervenience, should be faithful to the current debate.

or on the problematic experience of dissatisfaction, when approaching the end of a paper?). And are there reasons to assume that at least specific aspects of mentality fail to supervene?

Current literature on supervenience and on mental causation quite generally distinguishes between "intentional" (representational) mental properties and "qualitative"[27] (phenomenological) mental properties: "intentional" properties are often recognised as supervenient on physical properties, while "qualitative" properties are not. If *qualia*[28]—as they are called—fail to supervene or exhibit a weaker kind of dependence, perhaps they should be considered epiphenomenal, lacking real causal efficacy and explanatory power with respect to human behaviour. On the other hand *qualia* and their coherently conceivable "inversions" could be undetectable, even if we knew all about neurophysiology: this would undermine supervenience also in its weak version.

Problems for mind-body supervenience also arise with certain kinds of intentional mental states, what are termed the "wide-content" states: in fact their supervenience base includes not only the intrinsic physical features of the individual entertaining them but also certain connections with a wider environment. I could believe that water (H_2O) is transparent, while my physical duplicate on Twin Earth could believe that "twater" (XYZ, observationally indistinguishable from H_2O) is transparent. It would then follow that the belief that water is transparent —and all wide-content states—do not supervene on physical states. Of course it is possible to modify the definition of supervenience in such a way that the subvenient base includes extrinsic, or relational, properties of the subject. And from this starting point a variety of issues concerning wide- and narrow-content mental states are being discussed as central both for philosophy of mind and for cognitive science.

---

[27] Very synthetically, intentional mental properties are expressed by propositional attitudes by means of *that*-clauses ("She believes that Singapore is an attractive city"), while qualitative mental properties correspond to *what it is like* to be in a mental state, and hence they characterise sensations, feelings, perceptions and perhaps thoughts as well. There are quite famous thought experiments regarding both intentional mental properties (e. g. "Twin Earth" by Putnam) and *qualia* ("inverted spectrum": two functionally identical individuals could both call "red" things that look to one the way things they both call "green" look to the other).

[28] About *qualia*, see the somehow classical paper by Thomas Nagel (1974): "what is it like" for a bat to perceive the world through the peculiar sensory system of "echolocation"? A computational theory about how the system makes spatial information available to the bat cannot provide an answer, according to Nagel.

Qualia and intentionality[29] take directly to subjectivity and consciousness, the ultimate barrier for the understanding of mentality and for a comprehensive explanation of human beings (and perhaps of other organisms or "systems"). The scientific progress in investigating the neuronal correlate of consciousness and in developing cognitive models of it does not seem to help in answering central questions such as: what is it like to be conscious? how does it differ from not being conscious? why should a physical system with a specific architecture produce intentionality and subjectivity? The present paper could have started from questions like these; on the contrary, we did not even mention a possible distinction between conscious and unconscious mental states, and did not confront the central issue of subjectivity, in particular as a source of normative constraints within epistemological and moral concerns. The guiding line of supervenience apparently took us away from subjectivity and from a proper recognition of the heterogeneity of the mental with respect to the physical. Many efforts have been (and are currently being) dedicated to introducing "anomalies" within a materialist substance metaphysics and to "manipulating" the definition of supervenience in order to reconcile what could be irreconcilable (significant dependence with non-reducibility). Perhaps a divergent research direction could be more fruitful, reconsidering and taking seriously common-sense reports about conscious experiences and meta-reflections about them. After all, it is on such a basis that it is possible to distinguish between two very different sides of what is it like to be a human being, and to state the mind-body problem itself. According to Nagel,

> The success of a particular form of objectivity in expanding our grasp of some aspects of reality may tempt us to apply the same methods in areas where they will not work, either because those areas require a new kind of objectivity or because they are in some respect irreducibly subjective. The failure to recognize these limits produces various kinds of objective obstinacy—most notably

---

[29] There is an overlap between intentionality, definable as the feature of the mind of directing itself at objects or states of affairs in the world, and consciousness, but they are not coextensive. In fact, there are intentional states that are not conscious (like a belief of a sleeping person) and conscious states that are not intentional (like a feeling of fear without an object to be afraid of). About the relation between intentionality and consciousness, mostly kept separated by cognitive science and philosophy of mind, see Crane 2001. The book recalls Brentano's thesis about the intentionality of mental phenomena and defends emergentism as a form of true dualism.

reductive analyses of one type of thing in terms that are taken from the objective understanding of another[30].

Both an "objective" and a "subjective" point of view are apparently required to cover the still mysterious territory of mentality, where we can view ourselves from outside—tracing a naturalistic and unavoidably reductive picture of how we work—and where we can understand what it is like to have conscious experiences only by assuming the point of view of subjects capable of having them. "Double vision is the fate of creatures with a glimpse of the view *sub specie aeternitatis*"[31] and it could also be that no single—complex but consistent—view is reached or even reachable, that we simply have to acknowledge the existence of an "explanatory gap".

In conclusion, to explain mental properties as natural phenomena[32], for instance through the conceptual tools of evolutionary psychology, could produce their extensive assimilation among the adaptive features of human organisms. In this case, what would be the value of the very arguments supporting the evolutionary point of view, since selection could have favoured irrationality and even false beliefs? It seems that neither the theory of evolution nor any other physicalist theory could give a complete account of rationality, since any such theory would be unable to explain itself: we—both reductionists and antireductionists—unavoidably assume the normative attitude of defending the epistemic value of such theories and of assigning them an independent validity.[33]

References

Beakley, B. and Ludlov P. (eds): 1992, *The Philosophy of Mind*. Cambridge (USA): The MIT Press.
Beckermann, A., Flohr H., and Kim J. (eds.): 1992, *Emergence or Reduction?: Prospects for Nonreductive Materialism.*. Edited by R. Posner and G. Meggle. Foundations of Communication and Cognition. Berlin: Walter de Gruyter.

---

[30] See Nagel 1986, especially Ch. V ("Knowledge"), par. 6 ("Double Vision"), from where these lines are extracted.

[31] *Ibidem.*

[32] For a more detailed discussion of the risks of naturalism, due to its intrinsic circularity, see Montecucco 2000: the paper argues that the "naturalized" epistemologist, with his/her conjectures, invalidates the same cognitive tools used to support them.

[33] Nagel 1997 develops this conclusive point.

Beckermann, A.: 1992, "Supervenience, Emergence and Reduction", in, A. Beckermann, H. Flohr, and J. Kim (eds.): 1992, pp. 94-118.

Braddon-Mitchell, D. and Jackson F.: 1996, *Philosophy of Mind and Cognition.* Oxford—Cambridge (USA): Blackwell Publishers.

Broad, C. D.: 1925, *The Mind and its Place in Nature.* London: Routledge and Kegan Paul.

Clark, A.: 2001, *Mindware. An Introduction to the Philosophy of Cognitive Science.* New York—Oxford: Oxford University Press.

Cordeschi, R.: 2002, *The Discovery of the Artificial. Behaviour, Mind and Machines Before and Beyond Cybernetics.* Dordrecht: Kluwer Academic Publishers.

Crane, T.: 2001, *Elements of Mind. An Introduction to the Philosophy of Mind,* Oxford: Oxford University Press.

Davidson, D.: 1970, "Mental Events", in L. Foster and J.W. Swanson (eds.), *Experience and Theory.* Amherst: University of Massachusetts Press, pp. 79-101. Reprinted in Davidson, D.: 1980, *Essays on Actions and Events,* Oxford: Blackwell, and in Beakley and Ludlow 1992, pp. 137-149.

— 1987, "Problems in the explanation of action", in P. Pettit, R. Sylvan, J. Norman (eds.): 1987, *Metaphysics and Morality.* Oxford: Blackwell, pp. 35-49.

— 1992, "Thinking causes", in J. Heil and A. Mele (eds.): 1993, *Mental Causation.* Oxford: Clarendon Press.

— 1995, "Could there be a science of rationality?", in *International Journal of Philosophical Studies* 3, 1, pp. 1-16.

Haugeland, J.: 1997, *Mind Design II.* Cambridge, MA: The MIT Press.

Holland, J. H.: 1975, *Adaptation in Natural and Artificial Systems.* Ann Arbor, MI: University of Michingan Press. 2nd ed.: MIT Press, 1992.

Kim, J.: 1982, "Psychophysical Supervenience as a Mind-Body Theory", in *Cognition and Brain Theory* 5 (2), pp. 129-147.

— 1984, "Concepts of Supervenience", in *Philosophy and Phenomenological Research* 45, pp. 153-176. Reprinted in Kim 1993, pp. 53-78.

— 1987, "'Strong' and 'Global' Supervenience Revisited", in *Philosophy and Phenomenological Research* 48, pp. 315-326. Reprinted in Kim 1993, pp. 79-91.

— 1989, "The Myth of non-reductive Materialism". *Proceedings and Addresses of the American Philosophical Association* 63: 31-47. Reprinted in Kim 1993, pp. 265-284.

— 1990, "Supervenience as a Philosophical Concept". *Metaphilosophy* 21. pp. 1-27. Reprinted in Kim 1993, pp.131-160.

— 1993, *Supervenience and Mind.* Cambridge University Press.

McLaughlin, B. P.: 1992, "The Rise and Fall of British Emergentism", in A. Beckermann, H. Flohr and J. Kim (eds.), *Emergence or Reduction? Prospects for Nonreductive Materialism.* Berlin: De Grutyer, pp. 49-93.

Montecucco, L.: 2000, "Naturalismo e normatività", in *Epistemologia* 23 (1), pp. 147- 167.

Nagel, E.: 1961, *The Structure of Science.* Harcourt, Brace & World, Inc.

Nagel, T.: 1974, "What is it Like to be a Bat?", in *Philosophical Review* 4, pp. 435-450.
— 1986, *The View from Nowhere*. New York: Oxford University Press.
— 1997, *The Last World*. New York: Oxford University Press.
Nicolis, G. and Prigogine, I.: 1989, *Exploring Complexity: An Introduction*. New York: W.H. Freeman and Company.
Putnam, H.: 1967, "The Nature of Mental States", in H. Putnam: 1975, *Philosophical Papers, Vol. 2, Mind Language and Reality*. Cambridge: Cambridge University Press, pp. 429-440.
Stephan, A.: 1992, "Emergence—A Systematic View on its Historical Facets", in A. Beckermann, H. Flohr and J. Kim (eds.), *Emergence or Reduction? Prospects for Nonreductive Materialism*. Berlin: De Grutyer, pp. 25-48.
Villanueva, E. (ed.): 1993, *Naturalism and Normativity*. Atascadero (CA): Ridgeview Publishing Company.

13. From Complexity Levels to the Separate Soul

Giuseppe Del Re

University of Neaples, "Federico II"

1. From complexity to ontology

The word "complexity" has been applied, at least since the seminal work of Ludwig von Bertalanffy[1], to the study of the emergent properties of a system that is a whole consisting of interacting parts playing specific roles. Since there still seems to be much confusion about the corresponding concept, especially in connection with ontology, we shall devote the first part of this paper to its ontological significance, as well as to the related concepts of emergence and complexity levels. This will allow us to reach a point where life and the soul can be looked at in the light of the current universe of concepts of the natural and human sciences.

1.1. Perspectives of complexity

Within the sciences of nature—particularly physics, chemistry and biology—two main tendencies with regard to complexity appear to coexist. We shall call them the "process perspective" and the "systems perspective". As will be seen, they are not different approaches to the same problem, but genuinely different views of what complexity is. As a matter of fact, they are

---

[1] Bertalanffy 1962. See also the introduction by E. Agazzi to a collection of Italian translations of the source papers on this subject: E. Agazzi 1978.

related to the currently emerging clash between physics and biology as models of scientific inquiry and philosophy of science.

The process perspective is usually held by the physicists. In a nutshell, it considers chains of processes, each having more than one possible end result, and discusses how a single initial event can give rise to an intricate pattern of change. Studies based on this sort of problem are frequent in physics as well as in the theory of information flow (communication). On etymological grounds, the term 'complicated' is more appropriate than 'complex' here. This consideration is important because, on the one hand, the process perspective can make a significant contribution to today's scientific thought (or philosophy of nature), e.g. in connection with the theory of natural selection and decision theory. On the other hand, use of the term 'complexity' in connection with complicated networks of processes is liable to distract attention from complexity as a bridge between reductionistic and holistic approaches to the understanding of systems; in which case the term in question does play a unique role.

In fact, the systems perspective of complexity is held mainly by the biologists, who are confronted with the reality of organisms. It denotes an aspect of reality—the existence of entities consisting of distinct parts ('systems') having a measure of unitary character (particularly living beings)—which mechanistic approaches and the most popular philosophy of science have largely ignored.

We shall place ourselves in the systems perspective, and define complexity as *the qualifying property of a system whose parts combine to produce a whole exhibiting properties qualitatively different from those of the parts*. In short, in the systems perspective, complexity is defined in terms of the age-old problem of "the whole and its parts". This definition seems to fit the more original part of the views of the pioneers—Wiener, Bertalanffy, Morin, Prigogine and others—even though they seldom distinguish between systems and processes.

1.2. About systems and living beings

We normally speak of wholes or units as if everything were clear with regard to the concepts involved. But a few examples will make many difficulties apparent.

Take the solar system. Is it a whole, and is it a whole in the same sense as a living being? The answer is not easy. It is a well-defined part of the universe, very weakly interacting with the rest and formed of (comparatively) strongly interacting parts, the planets, the asteroids, the sun; but its properties are additive up to minor corrections, and, what is more, they are passive. They are 'additive' in the sense that the global properties of the whole, e.g. its mass and its angular momentum, are just sums of the corresponding properties of the parts; they are 'passive', for the solar system does not operate—i.e. act with exchange of information—in any way on its environment. It only generates a gravitational field which tells its environment that it is there; but it is wholly indifferent to anything but catastrophic events in its close neighborhood. Nor does it depend on its environment for its conservation. Other systems (e.g. molecules)[2] exhibit properties that are qualitatively different from those of their parts, but fall in the same class as the solar system in that they are passive.

By contrast, consider a closed-loop control system (to be abbreviated CLCS), which is the simplest type of whole sharing its specific characteristics with living organisms[3]. A concrete example is a TV set. A CLCS is as certainly a system as is the solar system, for it is clearly distinct from its environment; but it is active, and that in many ways.

First, it has at least two different states, which we may call its off-state and its on-state. When it is off, it is a collection of components arranged in a certain configuration and held together by screws, welding, etc. Given a disturbance it will behave completely passively. In the on-state (in a TV-set either 'stand-by' or fully operative), it acquires new properties and becomes *active*, because those new properties include special abilities or *faculties*, particularly that of selecting and processing signals of a certain kind arriving at its input terminals, and producing signals of the same or another sort at its output terminals (e.g. VHF electromagnetic waves as input and images as output).

Secondly, the ability to process information is not the sum of the properties of the single components, although each of them plays a role in the task or in subsidiary operations. Not less important, the components operate in the way they do because in the on-state of the whole they perform their operations by changing their states according to the states of the other components (for example, each transistor operates with terminal voltages which depend on the currents passing through the components connected to it). A CLCS thus appears to be 'active' both internally and externally, because it is not a CLCS if its parts do not have properties that are only present when

---

[2] Del Re 1987.

[3] Illustrations of this can be found, for example, in textbooks of neurophysiology.

they exchange information (in the form, say, of an electric current) with one another.

Thirdly, a CLCS depends on its environment in at least two respects: it needs as a source a specific form of energy (in the TV-set example, electricity), and it adjusts to the environmental conditions by appropriate feedback loops, which modify its states according to changes in external parameters (temperature, humidity, etc.), so as to make the significant features of its responses to input signals independent of those parameters (e.g. produce images of the same quality and definition even if the average intensity of the input signals or the external temperature change).

The above points can be found in contemporary textbooks on systems science and engineering; but such is not the case for their all-important ontological implication, namely that a CLCS is an entity made up of a whole consisting of many parts that have internal and external properties whose nature cannot be predicted from the properties of the parts in isolation. This "qualitative unpredictability" holds even when, as in the simpler cases of engineering science, the properties of a whole can be represented by linear combinations of the properties of the parts *in situ*; it is only if the nature of a property of the whole is known, that the value of one or more measurable properties can be predicted.

We emphasize that a typical artificial CLCS is capable of performing its function under conditions, e.g. the temperature and the mean amplitude of the input signal, that are subject to random fluctuations, albeit within a certain range. The same sort of behavior, in a far richer form, is exhibited by a living being. It is therefore legitimate to say that a living being is a highly sophisticated closed-loop control system. What about its parts? Well, organ transplantations have proven that the organs of a living being are perfectly capable of functioning as separate entities, provided the necessary conditions are realized (temperature, supply of nutrients, etc.). Therefore we can identify (as did Aristotle more than two thousand years ago) the organs as actual (not just virtual) parts of a living organism, parts whose disposition and connections constitute the second matter which life (the "soul") actualizes[4]; and we also know that no characteristic property of a living being can be said to abide in individual organs. Attempts have been made at

---

[4] Aristotle, *De Anima* II, 412a, Engl. transl. by J. A. Smith in Aristotle 1984. Note that we use "soul" to denote the soul according to Aristotle, reserving the word 'psyche' for that part of it that is the object of psychology.

conceptually reducing a living being to a brain, but biology and the medical sciences have shown that this is a simplification having no root in reality.

In a completely integrated CLCS, particularly a living being, the parts are active, and depend on one another in order to retain the properties that make them parts of the given whole. Due to the chemical nature of physiological activity, in living beings this means that any part will decay to become a nonliving object if it is unable to receive appropriate signals from the other parts or from sophisticated technical devices that simulate them.

1.3. Complexity as an autonomous field of inquiry

The above discussion allows us to claim that complexity is a field of scientific inquiry. The branch of science that studies it (to be called 'complexity science', or 'complexity' for short) may be defined as *the collection of procedures and the universe of principles and concepts required for understanding by which rules and conditions the parts of a system may or may not give rise to a unit endowed with properties that cannot be traced back to those of the parts*. Although we place ourselves in a systems perspective, this definition also covers the process perspective, because the processes of which the coordinated activity of a CLCS consists are certainly part of the subject matter of complexity studies.

Complexity science is distinct from other branches of science in that it has its own object and its own program. It is expected to have its own system of axioms and laws (although much progress is needed in this direction), possibly taken over from other fields of inquiry, but developed and applied independently of the latter in accordance with the specific aim of understanding complexity.

In addition, complexity is a meta-disciplinary field somewhat like environmental science, in that its data consist in information already processed by other disciplines. Even Bertalanffy's general theory of systems, which provides the theoretical framework that can be considered proper to complexity studies, owes many of its results to physics and applied mathematics.

1.4. Emergence

The appearance of new properties (or new order) in an ensemble of parts is often called emergence. As is well known, this concept has given rise to some suspicion, since it is very popular in theories of life and of knowledge that have a strong ideological bias. In our opinion those suspicions are not justified, and emergence is a concept needed in any study of complexity. The justification of our statement is as follows.

First, all scientific discoveries and theories support the axiom "all that is becomes", provided the copula "is" refers to material entities or entities that require material support. This means that all things have a past, a present, and a future as such entities or as different ones (within the limits of mass and energy conservation).

Second, becoming in the universe generates new, more complex entities and destroys other entities. For example, according to highly reasonable theories, stars are formed from gaseous nebulae and lose much of their matter as radiation into space. Living beings grow by assimilating nonliving matter and become old (probably) as a result of built-in processes that reduce their efficiency. That there should be loss of order and coherence is easily understandable, and sanctioned by the second principle of thermodynamics. On the other hand, the appearance of new ordered structures poses a problem, which becomes a cause of wonder when it is not just the generation of individuals of a given species, although even the multiplication of a species is in apparent contrast with the second principle of thermodynamics. Ilya Prigogine's pioneering studies have shown that order may and actually does appear out of disorder. In the famous example given by Prigogine, Bénard's structures, there is an external ordering factor, a constant uniform temperature field. In the possible spontaneous origin of life there seems to be no ordering field, but a careful analysis shows that chemical *selection rules* might have guided the process, so that the adjective 'spontaneous' only means that a random event could be the beginning of a sequence of processes to which only certain end results were open. This is close to the idea of Darwinian selection, and implies that, since there exists in the universe a basic drive to change, certain events, including the spontaneous appearance of life, are in the order of things[5].

The formation of ordered structures from chaotic (or less ordered) ensembles by processes controlled by selection rules is but the actualization of latent (potential) properties of the initial ensemble: selection rules focus a sequence of processes more and more finely, so that the becoming of the given ensemble leads to one or just a few alternative results.

---

[5] A more detailed discussion is given in: Del Re 2000a, chh. 3, 6 and passim. See also Del Re 1994.

A simple example[6] will make the point clearer. Let us take the extreme case of a chain of processes (steps) that are all stochastic. Suppose our chain consists essentially of two steps, with the following tree: A branches out to B, C, D with probabilities of 20, 30, and 50 percent; B branches out to E, F, and G with the same probabilities; C branches out to E′, F′, G′ with probabilities of 15, 25, and 60 percent; D branches out to E″, F″, G″ with probabilities of 5, 15, and 80 percent. The overall probabilities are given by the following table:

E F G E′ F′ G′ E″ F″ G″
3 5 12 4.5 7.5 18 2.5 7.5 40

Evidently, the sequence A-D-G″ is greatly favored. The higher probabilities of the outcomes D and G″ mean the following: "D is more likely to be the event following A than the two others; and, *if D happens to be the actual outcome* of the first step, then G″ is most likely to be the outcome of the second step". The interesting fact is that, when we speak of probabilities, we usually think of repeated trials. But in each step it is the actual outcome that matters, and that is determined in succession by the actual outcome at each step. One may summarize this by saying that each process is 'result oriented', an expression which also alludes to the fact that in stochastic processes occurring in nature the probabilities and the very events that can follow a given event, say D, often depend on the effect of D on other processes going on in the environment, so that the probabilities associated with the entire process cannot be predicted at the beginning.

The above example shows that in general all sequences of processes, even all stochastic processes, will either be lost in chaos or converge toward a unique final result. The overall mechanism by which this final result is realized may be called "emergence".

The definition of emergence is thus given in what may be called the "how-mode", because no light is thrown on what emergence is. This is why it sounds as if it were a great new discovery of modern science. If one moves to the "what mode", however, one finds that emergence is contained in Aristotle's solution to the problem of becoming, with the correlated concepts of potentiality and actuality: it is the *actualization* of potentialities of matter that lead to a higher degree of order, and spontaneous emergence is the same

---

[6] Del Re 2000a, ch. 4, p. 101.

actualization that corresponds to a built-in tendency of matter—simply what Aristotle would have called its *physis*, its nature.

Two points should be underlined. First, according to Aristotle change in general is the actualization of potentialities, whereas emergence refers only to change corresponding to augmented order; second, emergence never concerns form in opposition to pure matter, but the actualization of new features from a material, a *materia secunda*, in accordance with Aristotle's remark that

> In every class of things, as in nature as a whole, we find two factors involved, a matter which is potentially all the particulars included in the class, a cause that is productive in the sense that it makes them all (the latter standing to the former as, e.g., an art to its material)[7].

The term 'art', of course, is here the translation of the word *techne*, and means the manufacturing or fabrication of an object. In the case of emergence a sort of spontaneity is understood in contemporary thought, but this does not reduce the validity of Aristotle's remark, as Aquinas pointed out:

> In nullo enim alio modo natura ab arte videtur differre, nisi quia natura est principium intrinsecum, et ars est principium extrinsecum. Si enim ars factiva navis esset intrinseca ligno, facta fuisset navis a natura, sicut modo fit ab arte...; natura nihil est aliud quam ratio cujusdam artis, scilicet divinae, qua ipsae res moventur ad finem determinatum, sicut si artifex factor navis posset lignis tribuere, quod ex se ipsis moverentur ad navis formam inducendam[8].—It seems that nature does not differ from a craft, but because nature is an intrinsic principle, whereas a craft is an extrinsic principle. If in fact the skills of ship-making were intrinsic to wood, a ship would be made by nature precisely as it is made by craft. ... Nature is nothing but the essence of a certain craft, divine of course, by which things evolve toward a determined end, as if the designer and builder of a ship could confer to her pieces the ability to move spontaneously to yield the shape of the ship.

---

[7] Aristotle, *De Anima,* 3.5, in Aristotle 1984, I, p. 641.

[8] Aristotle, *Physic,* LII, l.XIV.

1.5. Levels of complexity and Seinsschichten

As mentioned, although it is part of science, complexity is of great import for ontology, indeed it is as it were the database for that *Kategorienlehre* (science of categories) which Nicolai Hartmann rightly considered a major task of ontology[9]. A great merit of Hartmann was his realization that ontology is based on our experience of the world, and thus can only be a science—with due respect to Kant—if it takes into account the natural sciences. Moreover, he realized and developed an essential point in Aristotle's matter-form antinomy, namely that it is not limited to prime matter: there is a many-runged ladder of being:

> If you take the extremes, matter is pure matter and the essence is pure definition; but the bodies intermediate between the two are related to each in proportion as they are near to either[10].[10] The notion of "layers of being" (*Seinsschichten*), the key notion in Hartmann's ontology, is fully in line with this remark, which may be seen as a hierarchy of classes of objects whose general characteristics become richer and richer, each class requiring new ontological categories.

In connection with complexity, our application of Aristotle's and Hartmann's idea is as follows. At each layer of being, *which we shall also call a "level of complexity"*, there are objects that in themselves are completely determined, their potential properties being uniquely determined by their nature; we shall call them the "elementary objects" of that complexity level. Any one of these objects cannot be in itself a second matter for anything, but a collection of them may have undetermined potentialities. If this is the case let us consider the simplest new wholes having a number of them as parts; these wholes will require new categories because, if those applying to their parts were sufficient, they would not be new units. The new categories thus define a new layer of being.

Along this line we can also speak of an inherent complexity-level hierarchy proper to a given entity, which can also be called the *levels of*

---

[9] Hartmann has been called a genius. Some scholars complain that he did not include in his ontology spiritual entities subsisting *per se*; judging from his masterpiece, Hartmann 1964, it seems to us that the problem of categories such as he formulates it simply does not apply to spiritual substances.

[10] Aristotle, *Meteor*, IV-12, 390, 4-8, Engl. transl. by F. H. Fobes, in Aristotle 1984, I, p. 626.

reality proper to that object. We give briefly the argument leading to this notion; we shall return to it later.

The starting point of experimental science and technology, whatever certain philosophical schools may claim, is a strong realistic commitment, namely the belief that our sensations, provided that they be critically assessed by reason, inform us about a reality existing independently of us. Now, every object—except, possibly, genuine elementary particles like the electron—can be described as a collection of elementary objects belonging to different layers of reality or levels of complexity.

For example, a vividly colored fish of the Great Barrier Reef may be described as a special configuration of elementary particles (lowest complexity level), or of electrons and nuclei (higher complexity level); but it is also an organized ensemble of living cells (much higher complexity level), and a collection of organs acting together with a result-oriented organization (still higher complexity level), as well as being a colored fish (highest complexity level): *the potentialities for further information present in each description of its reality decrease as the degree of complexity under consideration increases*, since each description corresponds to a particular separation between matter (*materia secunda*) and a 'residual' form. If the residual form is ignored, then it can be said that the fish is being described at the *level of reality* of the objects forming what has been treated as matter, e.g. interacting cells. So those properties are ignored that *emerge* at a higher level because, out of all possible interactions of the cells, certain specific ones are actualized in the given fish. Indeed, the fact that it is that particular *living* fish and not any other object or being that could have been constituted with the same elementary particles or the same cells is only evident at the highest level of complexity. The latter corresponds to the highest level of reality for the fish, indeed to its proper layer of being, in the sense that whatever properties pertain to the lower levels are completed by the properties specific of a living being and by the individuation as "this particular living being", as Aquinas would have said. At the level of complexity proper of the fish, no room is left for the emergence of new properties, except those programmed in the built-in project typical of living beings.

The complexity version of realism, combined with the notion of layers of being, thus provides a foundation for an ontology that takes the most recent advances of science into account: after all, it is the nature and relations of the elementary particles making up a body that lie at the bottom of its ladder of reality levels; and that nature and those relations are the same for all objects. However, it is essential that the difference between the how-mode and the what-mode should be kept clear. The following questions are an aid to this end:

- *how-mode*: how does a property of a whole arise from a cooperation of its parts?
- *what-mode*: to what can the properties of a whole be traced that cannot be ascribed to individual parts or groups of parts?

1.6. Key concepts: structure, organization, information

The fact that complex systems have properties that do not arise from a different number and sort of parts is already evident in the case of molecules, which are characterized not only by the species and number of their atoms, but by their "structure"[11]. Since, at least according to scientific criteria, a molecule is a well defined entity that behaves as a whole, it seems clear that structure is a characteristic that falls within the realm of complexity. Molecules, however, belong to a layer of being where entities subsisting *per se* are stable systems; and, as we have already seen (though not emphasized) in connection with control systems, the most significant complex systems are active ones. For the latter a conceptual step beyond structure must be taken. Structure becomes a marginal aspect of unity, and its role is played by 'organization', which is typical of stationary systems out of equilibrium, and in particular characterizes living beings.

A unifying approach to complexity can be found by appealing to information science. The latter makes it possible (though only in principle) to quantify the essence of an entity such as we know it (as *quidditas*) by evaluating the length of the shortest string that describes it completely in a suitable language. For an intuitive idea of this point, let us consider a collection of n objects. If the situation is such that the resulting system has no property of its own that is not the sum of those of the parts, the string l_o associated to it need only contain sums of properties of the parts. If, on the contrary, new properties emerge, not only is it necessary to add information about the arrangement and interaction of the parts, but the new properties must be described: hence a greater length of the string l expressing the quiddity of the system.

---

[11] Del Re 1998.

The main interest of the approach to complexity just outlined is that in it the way in which the parts interact to yield a unitary whole does not matter, nor need the parts and the whole be entities subsisting *per se*; they can be drawings, patterns of behavior, crowds, etc. It can be looked at in the very ancient and very modern sense of characteristics imparted to a suitable material to make it a specific object, and also—a point which will appear to be very important in the sequel of this paper—as *either* the record of a message imprinted on that material, *or* the very imprinting of the message, where by 'message' we mean any external action causing the receiving end to change its characteristics.

In other words, the term 'information' is used in two senses: (a) to denote the action of actualizing properties by imprinting, (b) to denote some aspect of an object that can be described in a language. A familiar analogy for both senses is recording on a compact disk. This notion is also suggested by Aristotle's favorite examples, the clay vase and the wax figurine. What about a living being, which changes all the time? Actually, nothing prevents the application of the same concept to entities whose shape and behavior change all the time, but then the nature of the underlying resident information is more elusive, unless a distinction is made between potential information stored in the genetic memory (the "nature") of the given being and actual information.

2. The nature of life[12]

The great contribution to science of the systems perspective of complexity has been the solution of the problem of life. Why are living beings so different from nonliving objects such that one is tempted to think that special principles and laws should be introduced for them? As already briefly recalled, the answer, mainly due to Prigogine and Bertalanffy, is that living beings are physical systems of a very special class, namely *stationary systems out of equilibrium*, more precisely closed-loop control systems. The philosophical implications of this view of life require a return to Aristotle's metaphysics with the mediation of Nicolai Hartmann's ontology. The nature of the human soul can therefore be reviewed in the light of complexity and information; indeed, even the famous stumbling block of the human soul as a

---

[12] Del Re 1997. See Del Re 2000.

"substantial form"can be reformulated in terms of an analogy with a control system that shows how the genius of Aquinas had anticipated concepts that have become familiar only in the last century. We shall examine these points, using double quotes to denote Aristotle's notion of soul, which is much broader in scope that the current meaning of 'soul'.

2.1. Organization

As mentioned, the key concept that appears to capture the qualifying properties of living is *organization*. As in the case of many difficult words, this term is used in many questionable ways, but the name "organization" given to many enterprises that sell services, e.g. airlines, justifies the claim that even in the popular mind it implies a dynamic cooperation of parts aimed at performing a given task. A sales "organization", for example, is a unit composed of many elements, and is expected to perform a specific task as a unit in a variable environment. What task? You may call it ensuring self-survival, or more generally protecting its own identity in the face of a changing environment, by adjusting all the time to external (and internal) fluctuations and disturbances. That is to say, a genuine sales "organization" will ensure as far as possible the supply to all customers of the sorts of goods, prices and terms of delivery in its catalogue on its entire network regardless of difficulties of all kinds; and the fact that this requires an organized activity is precisely what makes it a sales "organization".

The double quotes used above serve to remind the reader that the name 'organization' is applied metaphorically in a socio-economic context, to denote an entity that has a material support (a structure consisting of staff, offices, trucks, etc.) *and* an organized activity. Properly speaking, on the other hand, it denotes, in the how-mode, a special kind of coordinated activity and, in the what-mode, a principle, never a system (i.e., not a material 'substance'). As happens for many substantives of the same kind (in particular 'information'), it denotes both a fact (to be organized) and an action (organizing). Usually, this is not a cause of confusion. Confusion may arise, instead, from the fact that people often speak of 'organization' meaning 'structure'. We have associated to the former the idea of a unitary targeted activity, because this also seems to be implicit in current use (as with sales organizations). When useful, longer expressions, such as "coordinated result-oriented activity of the parts", may be necessary. Note that the latter expression clearly shows the closeness of the concept we are speaking of to Aristotle's *entelécheia*.

Organization in the sense here adopted appears to be a necessary condition for the result-oriented behavior of a system acting as a whole and preserving its identity in a variable context. It is precisely that kind of interdependence of the parts of a whole that makes it possible for a unitary system to adjust its behavior and its internal activity to changes in the environment, perceived as external stimuli or input signals, as well as to internal changes, so as to ensure preservation of its identity or execution of a pre-established program. Typical examples are the self-defense and immunity responses of a living being and the automatic route corrections of space probes. The qualification 'result-oriented' refers to interacting parts all working coherently toward one or several specific ends—say, self-preservation, propagation of the species, etc.

Organization may be seen as a high-quality sort of "information", inasmuch as it actualizes some or all of the properties to which the parts might give rise if driven, so to speak, to do so. (We have mentioned above the universal drive to change that science simply considers a fundamental aspect of material reality, indeed, a condition for its intelligibility).

2.2. Levels of reality of a living being

In order to proceed toward a discussion of the soul in the light of complexity levels as levels of reality, let us first of all emphasize that in living beings organization is the result of elementary processes at a variety of levels, forming a sort of hierarchical scale, as already shown in the example of sec. 1.5. Among such levels the following are relevant to this study:

- the level which takes atoms as the ultimate building blocks of matter, where the transmission of biological signals of all kinds appears as a extremely intricate set of interdependent physico-chemical processes[13];
- the level at which the "elementary objects" are the enzymes and other bio-macromolecules;
- the level where the simplest units introduced to explain facts are the cells;
- the level at which tissues and organs are the ultimate parts to be considered;

---

[13] The term 'signal' is used here in a broad sense, and includes not only nervous signals, but processes such as the transport of glucose to the cells.

- the level at which the organized system of the organs is studied independently of its material substrate.

Each of these levels is characterized by a particular matter-form pair, as explained by Aristotle (see sec. 1.5), in the sense that the ensemble of the elementary objects is treated as matter, and the potentialities of the matter actualized in the given living being are treated as form. Since what we consider in our knowledge is form, whenever we prescind from the matter characteristic of a given level we are left with an object of study that is the correlated form regardless of ontological considerations; that object tells us what the given being is.

2.3 The "soul", i.e. the soul as life principle[14]

We have thus reached the conclusion that, in the complexity-level approach, a fundamental characteristic of a living being is that, at the topmost level, it appears as an entity endowed with properties and capable of an activity which cannot be attributed to individual 'components' or organs, nor to specific physiological processes. More precisely, one could imagine that the transition from the layer of being of nonliving systems to that of living beings begins from a body with organs that are fully operative but which do not operate coherently for specific ends, as they should for life to be present. Then the form correlated to such an ideal 'mere' body would be whatever must be added to it to make the relevant being alive, with all the faculties that are characteristic of life.

At this point Aristotle's definition of the "soul" acquires a clear-cut meaning in the context of complexity:

the "soul" is the first level of operation of a natural body having life potentially in it. The body so described is a body formed by organs. ... If we have to give a general formula applicable to all kinds of "soul", we must describe it as the first level of operation of a natural body capable of organized activity[15].

---

[14] We remind the reader that we use "soul" to denote the soul according to Aristotle, reserving the word 'psyche' for the object of psychology.

[15] Aristotle *De Anima*, II, 412a. In Aristotle 1984, Smith interpreted this passage as: *the soul is the first grade of actuality of a natural body having life potentially in it. The body so described is a body which is organized. ... If we have to give a general formula applicable to*

The "soul" of any living being whatsoever is thus referred to a material support which has become alive because its parts exchange messages with one another so as to perform a coordinated activity in view of certain ends. For the modern mind this sounds quite different from saying that the "soul" is *something*. That it is not has been discussed by so many philosophers that we shall confine ourselves to reminding the reader that, for example, a computer program is 'something' unitary and well defined whose nature is independent of the support on which it is recorded, even though it consists of instructions recorded on that support.

Just as a computer program is not material (it has no mass, for example), so is the "soul" of any living being, except possibly man. We shall consider this latter point presently, for we want to discuss beforehand the question: is the psyche as psychology studies it the same as the "soul" according to Aristotle?

2.4 The psyche and the human soul

Strictly speaking, the question just formulated is undecidable, because no detailed definition of the 'body with operative organs' mentioned in Aristotle's definition has been given. We suggest that one could work the other way round, and define the psyche first, in which case the body is whatever remains—in Aristotelian language "whatever *materia secunda* the psyche is the 'form' of".

The psyche of a living being can be defined *grosso modo* as the given being as characterized by the properties emerging from the interaction of the elementary objects of the complexity level defined by neurophysiology. It is a new object of study, because its properties cannot be reduced to processes in the nervous system. It can be divided into parts (which of course are not treated as material objects), not only because no scientific study is possible without a measure of decomposition, but because its behavior can be described as that of a system capable of many states, more precisely *states of*

---

all kinds of soul, we must describe it as the first grade of actuality of a natural organized body. The main difference with our rendition comes from the Greek word *entelécheia*, translated as 'actualization' by Smith. It appears to mean here "the condition of a being or self-regulated control system actually operating as such", in full agreement with our translation. The word 'activity' used by other translators is better in certain contexts. A specialist's discussion can be found in Trendelenburg 1957, pp. 242-244.

consciousness[16]. This is why we have sciences called psychology and psychiatry.

Following Jaspers, we could include in a man's psyche also the intellect, and divide it into a noopsyche and a thymopsyche[17]. The question if what is currently called man's soul coincides with the psyche according to Jaspers remains open, because Jung's collective unconscious and archetypes might be considered as proper to the hardware of man's reality. Whatever the answer to that question may be, it remains undeniable that the frontier between the psyche and the body is not easily drawn. One might even doubt that any such frontier can be established even if the body were a material substrate with only the ordinary properties of a physical object—its weight, for example, is a property that can affect a person's behavior because of emotional reactions, say, to a fall. A way out of this difficulty is that whatever in the body is neither essential to life nor proper to the essence of man can be treated as an external cause of the responses it causes[18].

If we now take into account our whole discussion concerning organization, we can summarize the conclusions reached so far as follows: The psyche is the organization of a living organism inasmuch as it is responsible for its behavior as a whole. In terms of the ladder of complexity levels, it is what the being itself appears to be at the topmost level. Therefore, as long as its study is limited to that complexity level, it can be treated as an independent entity; but a full comprehension of the psyche requires consideration of all complexity levels, that is to say, of the full reality of the whole living being to which it belongs.

We insist that organization here is at the same time result-oriented coordinated activity and information resident in the body when the latter is activated (animated). In either capacity it requires a material support. Here is where the difficulties originated that were raised against Aquinas by the theologians of his time because of his acceptance of Aristotelian philosophy.

2.5. *Soul as* substantia

Let us now come to the major difficulty for the acceptance of the idea that man's "soul" is organized activity: man's ability to be aware that he knows.

---

[16] See e.g. Tart 1976.

[17] Jaspers 1913.

[18] The point that the soul is form with respect to an already partly informed body is emphasized by Aquinas in the *quaestio de anima* cited below, objection and response 14.

That this was difficult to reconcile with the notion of soul as form was already evident to Aristotle: although he did not think of organization as a coherent network of chemical processes as we do today, he considered all operations of the soul of an animal as the activity of a material support, and realized that self-consciousness is not by its nature the activity of anything. Many modern cognitivists have come to the same conclusion, although they formulate it by saying that self-consciousness does not seem susceptible of a definition such that a mechanism for it can be sought[19]. The same consideration led great neurophysiologists such as John C. Eccles to adopt a version of the old dualistic position, affirming that man's self is an entity separate from his body.

We shall not pause here to consider the reasons why the dualistic solution leads to more difficulties than it overcomes, because many of those reasons are connected with the Christian doctrine of Incarnation and Resurrection from the Dead, and here we are only concerned with philosophy of science[20]. We will rather grant that, (i) Aristotle's analysis is valid as far as it goes, (ii) the intellect (or self-consciousness), though a faculty of the soul of man, is not in itself dependent on a material substrate, and (iii) man's soul must be separable *in toto* from his body (as defined above)[21] and transferable to another body. A strictly scientific question is: granting the solution proposed by Aquinas of a form subsisting *per se*, can the standpoint of complexity make it compatible with science?

Before attempting to answer this question, let us point out that, except for radical determinists, no one denies that in man not only does the organization that is his "soul" interact with his body as if it were a single centre, but so does his intellect. One must therefore accept as a fact, unexplained in mechanistic terms but still a fact, the possibility that an entity that is *by its very nature nonmaterial* could exchange information with a suitable material system.

With these premises, let us examine Aquinas's attempt at reconciling Aristotle's concept of soul with the idea that it is transferable and not necessarily completely actualized in a natural body (as in the case of handicapped people)[22]. That argument can be extracted from a systematic

---

[19] See e.g. Chalmers 1996.

[20] A beautiful short review was given by Muldoon 1959, pp. 161ff.

[21] That is to say, as a biological substrate which plays the role of an interface between the soul and the sensible world.

[22] See Del Re 2000.

logical analysis of the statements in the *Prima quaestio disputata de anima*[23], and, as it seems, hinges entirely on the subtle distinction, to which we have already called attention in one or two places, between 'resident' information ('form') and in-formation (the action of imparting a 'form' or imprinting).

Do we have examples of either sort of information in the world of complexity? We do, and the most familiar ones are those CLCS's (closed-loop control systems, see sec. A.2) nobody would have thought of just a century ago, namely computers. A minimal computer consists of a main unit and two essential peripherals, a keyboard and a monitor. The main unit alone is not the actual computer, but only a potential one: for a computer is a signal processing unit, and if the essential peripherals are not connected to it the main unit cannot receive or emit any signals. Now suppose a monitor is connected to the main unit prior to loading any particular program. Depending on the hardware, the monitor will or will not show a message; but, if it does, it will say "no keyboard present". Skipping other details, let us pause on the 'metaphysical' significance of that message. It means that the main unit has checked its internal state and has found that it is not what it should be. The state in question can be identified with the form of the main unit with respect to what it is in its off-state, and therefore it must be concluded that *connection to a keyboard changes the state of the main unit*. Once a keyboard is connected, if the main unit contains a suitable built-in program, it will accept signals from it, but the main unit will have to send messages to the keyboard to determine various details, e.g. its repetition time. This may be translated into metaphysical language by saying that the main unit will have to impart a 'form' to the keyboard over and beyond the form it has of its own. To make a long story short, we reach the conclusion that, although the main unit exists *per se*, the actual computer is realized only if the essential peripherals are connected to it, and a mutual adjustment has taken place.

Let us now take our courage in both hands and forget that the main unit of a computer is by no means an "intellectual substance". Then we may consider an analogy with Aquinas's argument that runs as follows. Aquinas wrote:

Even if soul has a complete existence it doesn't follow that body is joined to it incidentally. For one reason, soul shares that very same

---

[23] The English translation to which we shall refer is Aquinas 1993, pp. 184-191. See Parenti OP 2000.

existence with body so that there is one existence of the whole composite; and for another, even though soul can subsist of itself, it doesn't have a complete specific nature, but body is joined to it to complete its nature[24].

With all respect to Aquinas, let us adapt the above statement to our computer example. It would then read (changes and additions in italics):

Even if *main unit* has a complete existence it doesn't follow that *the essential peripherals* are joined to it incidentally. For one reason, *the main unit* shares that very same existence *as an actual computer with its peripherals,* so that there is one existence of the whole composite; and for another, even though *the main unit* can subsist of itself, it doesn't have a complete specific nature (*species: "computer"*), but *the peripherals are* joined to it to complete its nature.

It seems to us that, despite its somewhat paradoxical and simplistic nature, the computer analogy shows that even by contemporary scientific standards Aquinas's solution to the separated-soul problem sounds completely reasonable, provided the system-perspective of complexity is duly taken into account.

2.6. A spiritual layer of being

A valuable brief history of philosophical thought reveals that for Hartmann it made no sense to look for an ultimate foundation of the "real world", whether a common act of being or a personal God[25]. We cannot take a stand on this statement, because we know Hartmann's thought directly only through his *Aufbau der realen Welt* and his brief review of his views on ontology[26]. But it seems to us that, by adopting the novel view that ontology must be the end point of the generalization of science, he could but stop at the boundary between whatever belongs to the sphere of our sensible *rationalizable* experience; for Hartmann makes an indispensable concept of what we have called "emergence", and thus excludes whatever we come in touch with by inner experience and intuition. In this sense, Aristotle is still

---

[24] Aquinas 1993, conclusion 1.

[25] Störig 1950,1969 vol. 2, p. 266.

[26] Hartmann 1942,1949.

one up on Hartmann, for he provides an argument showing if not proving that self-consciousness cannot be something emerging from matter.

Actually, it seems to us that Hartmann's *Seinsschichten* can be prolonged beyond the physical world precisely by using Aristotle's considerations as a point of support, provided it is admitted that in man there is an 'active mind' which, as the Philosopher would say, is not the coordinated activity (the *entelécheia*) of any material system. Then one must also admit that something nonmaterial by nature can act on entities that are supported by matter (the passive mind or the nervous system as information processor). Such an action, we repeat, cannot be regulated by scientific principles, because the scope of the latter does no extend beyond processes in matter; but it can be understood in terms of information and information transfer.

On this basis, we can call back to life a medieval discussion dear, among others, to Bonaventura da Bagnoregio and to Aquinas: the angels as possible pure spirits and their interaction with the sensible world[27]. The angels, creatures not in space-time, were attributed the power to inspire and illuminate human beings and, perhaps, living beings. This could be construed as the claim that certain immaterial entities could act on the *passive intellect* or the brain activity of a living being in the same way as the active mind normally does. Even the manifestations of angels in the form of men could be, so to speak, 'induced hallucinations'.

The example of angels illustrates the avowedly tentative claim that material (or matter-supported) and possible nonmaterial layers of being can be postulated without reducing the coherence and intelligibility of the whole picture. The interface between the spiritual layers of being is the psyche, since it can act on objects belonging to lower levels of reality (say, eyes) and can be acted upon by spiritual 'substances'.

We emphasize again that all we have just said is in need of a rigorous critical discussion; but it would seem that here is the direction in which work should be done to make it possible to add a few upper rungs to the ladder of being, specifically a set of 'pure spirit' layers of being. It would seem that what makes such an addition possible without loss of the unitary picture of an ontology based on a hierarchy of matter-form pairs, such as are implicit in contemporary science, is the relation between information and complexity.

In conclusion, the points here discussed can be reduced to the following list: a. the general problem of complexity science is to study; the relation between the how-mode to the what-mode of sensible reality; introducing

---

[27] See e.g. St. Bonaventura da Bagnoregio 1934.

levels of complexity that can be associated with Hartmann's levels of being; b. complexity science, therefore, places emphasis on 'resident' information, which coincides with 'form' in Aristotle's sense and is associated with the pertinent 'second matter'; c. the concept of information could make possible the extension to pure spirits (Aquinas's "intellectual substances") of a science of ontological categories consistent with the contents and criteria of reality of contemporary science[28].

References

Agazzi, E. (ed.): 1978, *I sistemi fra scienza e filosofia.* Torino: SEI.

Aquinas, T: 1993, *Selected Philosophical Writings*, Engl. transl. with an introduction and notes by Timothy McDermott. Oxford: Oxford U. Press.

Aristotle: 1984, *Complete Works*, ed. by J. Barnes. Princeton N.J.: Princeton/Bollinger.

Bertalanffy, L. v.: 1962, "General system theory—a critical review", *General systems* 1-20.

Bonaventura da Bagnoregio: 1934, "De existentia angelorum", in *Collationes in Hexaëmeron et Bonaventuriana quaedam selecta,* Ferd. Delorme OFM ed. Florentiae: Ad Claras Aquas, Tipografia del Collegio San Bonaventura, pp. 293-304.

Chalmers, D.: 1996, *The Conscious Mind: in Search of a Fundamental Theory.* Oxford: Oxford U. Press.

Del Re, G.: 1987, "The historical perspective and the specificity of chemistry", *Epistemologia* 10, pp. 231-240.

— 1994, "Information, Organization, Autopoiesis: From Molecules to Life", in B. Pullman (ed.): 1994, *The Emergence of Complexity in Mathematics, Physics, Chemistry, and Biology.* Vatican City: Pontificia Academia Scientiarum, pp. 277-293.

— 1997, "The question of the soul", *La Nuova Critica* (Rome), N.S. II, no. 30, pp. 75-98.

— 1998, "The Ontological Status of Molecular Structure", *Hyle, An International Journal for the Philosophy of Chemistry* (Karlsruhe) 4, pp. 1-23.

— 2000, "La questione dell'anima e la scienza di oggi" ("The question of the soul and today's science"), *Sapienza* (Naples, Italy) 53, pp. 383-418.

— 2000a, *The Cosmic Dance.* Philadelphia: Templeton Foundation Press.

Hartmann, N.: 1942,1949, *Neue Wege der Ontologie.* Stuttgart: Kohlhammer.

---

[28] An intuition of this point was presented a few years ago by a well known physicist, though in such a curious and philosophically naïve form that possibly valid points were enshrouded in a fog of imprecision and inconsistency: Tipler 1994.

—— 1964, *Der Aufbau der realen Welt.* Berlin: De Gruyter

Jaspers, K.: 1913, *Allgemeine Psychopathologie,* engl. transl. by J. Hoenig and M. W. Hamilton: 1997, *General Psychopathology.* Baltimore: Johns Hopkins University Press.

Muldoon, T. std: 1959, *Theologiae dogmaticae praelectiones. Volumen III, De Deo creante et elevante.* Roma: Officium Libri Catholici.

Parenti, S. OP: 2000, "Un testo di San Tommaso sull'anima forma del corpo" ("A text by Aquinas of the soul as 'form' of the body"), *Sapienza* (Naples, Italy) 53, pp. 354-381.

Störig, H. J.: 1950,1969, *Kleine Weltgeschichte der Philosophie.* Frankfurt a.M.: Fischer Taschenbuch Verlag.

Tart, C. A.: 1976, "The basic nature of altered states of consciousness: a systems approach", *Journ. Transp. Psychol.* 8, pp. 45-64.

Tipler ,F. J.: 1994, *The Physics of Immortality.* New York : Doubleday.

Trendelenburg, F. A.: 1957, *Aristotelis de anima libri tres.* Graz: Akad. Druck- u. Verlagsanstalt.

Index